AF584857

Pearson Australia
(a division of Pearson Australia Group Pty Ltd)
707 Collins Street, Melbourne, Victoria 3008
PO Box 23360, Melbourne, Victoria 8012
www.pearson.com.au

First published 2016 by Pearson Australia
2021 2020 2019 2018 2017
10 9 8 7 6 5 4 3 2 1

Publisher: Alicia Brown
Development Editor: Shirley Melissas
Project Manager: Shelly Wang
Production Manager: Elizabeth Gosman
Editors: Deborah Lombardo, Aptara
Designer: Anne Donald
Copyright & Pictures Editor: Samantha Russell-Tulip
Desktop Operators: Dave Doyle, Aptara
Illustrator: DiacriTech
Printed and bound in Australia by Pegasus Media & Logistics

9781488615085

Pearson Australia Group Pty Ltd ABN 40 004 245 943

Disclaimer/s
The selection of internet addresses (URLs) provided for this book/resource was valid at the time of publication and was chosen as being appropriate for use as a secondary education research tool. However, due to the dynamic nature of the internet, some addresses may have changed, may have ceased to exist since publication, or may inadvertently link to sites with content that could be considered offensive or inappropriate. While the authors and publisher regret any inconvenience this may cause readers, no responsibility for any such changes or unforeseeable errors can be accepted by either the authors or the publisher.

Some of the images used in *Pearson Science* might have associations with deceased Indigenous Australians. Please be aware that these images might cause sadness or distress in Aboriginal or Torres Strait Islander communities.

Practical activities:
All practical activities, including the illustrations, are provided as a guide only and the accuracy of such information cannot be guaranteed. Teachers must assess the appropriateness of an activity and take into account the experience of their students and facilities available. Additionally, all practical activities should be trialled before they are attempted with students and a risk assessment must be completed. All care should be taken and appropriate protective clothing and equipment should be worn when carrying out any practical activity. Although all practical activities have been written with safety in mind, Pearson Australia and the authors do not accept any responsibility for the information contained in or relating to the practical activities, and are not liable for any loss and/or injury arising from or sustained as a result of conducting any of the practical activities described in this book.

Acknowledgements
We would like to thank the following for permission to reproduce copyright material. The following abbreviations are used in this list: t = top, b = bottom, l = left, r = right, c = centre, (e) = ebook only.

Cover image: Science Photo Library/Gred Guenther

123RF: vitalii nesterchuk, p. 228l (e); reddz, p. 228r (e).

Alamy Stock Photo: Natural Visions/Alamy Stock Photo, p. 113r; birdpix, p. 55l.

Australian Bureau of Statistics: © Commonwealth of Australia. ABS, RP Data. Licensed under a Creative Commons Attribution 2.5 Australia license (CC BY 2.5 AU), p. 6.

Bureau of Meteorology: Based on Southern Oscillation Index (SOI) since 1876, http://www.bom.gov.au/climate/current/soi2.shtml. Bureau of Meteorology, © Commonwealth of Australia. Licensed under Creative Commons Attribution License (CC BY 4.0), p. 106.

Commonwealth of Australia: Department of Infrastructure, Transport, Regional Development and Local Government 2009, pp. 136b, 137, 138; Based on Road Deaths Australia 2008 Statistical Summary © Commonwealth of Australia 2009, p. 136t.

DK Images: Steve Gorton, p. 36bl; Colin Keates, pp. 35tr, 35cr, 35bl, 36tl, 36tr, 36br.

Getty Images: Strnadova Barbara, p. 60; Jason Edwards, p. 113l; Sovfoto, p. 43.

Office of Environment & Heritage NSW: Save Power Campaign, NSW Government, p. 230 (e).

Picture Media Pty Ltd: Reuters: 'Mobile phones distract drivers more than passengers' by Maggie Fox, ed. Cynthia Osterman, 1 December 2008, p. 134.

Science Photo Library: Wim Van Egmond, p. 10; Gred Guenther, p. I; Eckhard Slawik, p. 123; Carol & Mike Werner/Visuals Unlimited, Inc., p. 172b (e).

Shutterstock: Air Images, p. 134; alice-photo, p. 35cl; Marcel Clemens, p. 35tl; Digital Storm, p. 165; isak55, p. 25; Eric Isselee, p. 55r; Kateryna Kessariiska, p. 149; Oleksiy Mark, p. 228c (e); B.G. Smith, p. 35br.

The YGS Group: The Associated Press, 'Thirty years on, DNA tests free US man convicted of rape', The Age, 6 May 2010, p. 156.

Every effort has been made to trace and acknowledge copyright. However, should any infringement have occurred, the publishers tender their apologies and invite copyright owners to contact them.

PEARSON Australian Curriculum Writing and Development Team

Anna Bennett
Differentiation consultant, Victoria

Ian Bentley
Former Head of Science, VCE exam and trial exam writer, STEM investigation developer, Victoria

Christina Bliss
Science and senior Biology teacher, VCE assessor, Biology author and question writer, Victoria

Donna Chapman
Science laboratory technician, safety consultant, Victoria

Dr Warrick Clarke
Curriculum writer, Science communicator and Australian Post-Doctoral Research Fellow at UNSW, Author, New South Wales

Jacinta Devlin
Science and senior Physics teacher, author, Victoria

Julia Ferguson
Education Officer Earth Science Western Australia, author and reviewer, Western Australia

Bob Hoogendoorn
VCAA exam assessor in Chemistry, former senior Chemistry teacher exam-style question author, Victoria

Penny Lee
Science laboratory technician, safety consultant, Victoria

Louise Lennard
Head of Science, former industrial scientist, author and STEM investigation developer, Victoria

Greg Linstead
Former Head of Science, Education Department of WA curriculum writer, Deputy Chief Examiner, STAWA President, Author, Western Australia

Bryony Lowe
Director of Numeracy Improvement, former Head of Science, Region Teaching & Learning Coach, author and reviewer, Victoria

David Madden
Science Learning Area Manager at QCAA, former Head of Science, author and reviewer, Queensland

Fran Maher
Bioscience educator, Science teacher, former Bioscience researcher author, Victoria

Rochelle Manners
Science and Mathematics teacher, Teacher Companion coordinating author, Queensland

Shirley Melissas
Teacher librarian and author, Development editor, Victoria

Tamsin Moore
Science and senior Psychology teacher, author, Western Australia

Natalie Nejad
Head of Science, Science and Mathematics teacher, STEM investigation developer, Victoria

Malcolm Parsons
Education consultant, former teacher, author, Victoria

Greg Rickard
Teacher, former Head of Science, author, Victoria

Lana Salfinger
Teacher, Head of Science, IB Workshop leader in MYP Sciences author, Western Australia

Maggie Spenceley
Former teacher, curriculum writer Queensland Studies Authority, author, Queensland

Jim Sturgiss
Teacher, former coordinating analyst (NSW Department of Education) reporting NAPLAN, senior test designer for Essential Secondary Science Assessment (ESSA), thinking scientifically questions author, New South Wales

Craig Tilley
STAV trial exam coordinator, VCAA exam assessor, former senior Physics teacher, author and exam-style question writer, Victoria

Jo Watkins
Chief Executive Officer at Earth Science Western Australia, author and reviewer, Western Australia

Dr Trish Weekes
Science literacy consultant, New South Wales

Rebecca Wood
Science educator and tutor, author, Victoria

Contents

Access the worksheets for the Step up chapters in Biology, Chemistry and Physics on your eBook.

Answers are available in the teacher version of the eBook and on Productlink.

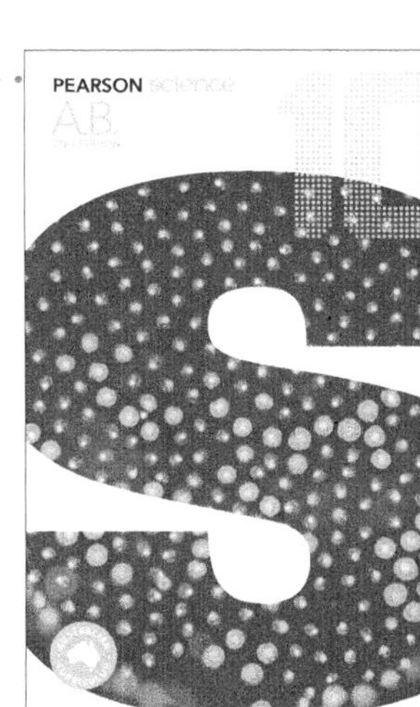

Go to your eBook to access the Step-up chapters.

How to use this book • ACTIVITY BOOK

Pearson Science 2nd edition Activity Book

An intuitive, self-paced approach to science education, which ensures every student has opportunities to practise, apply and extend their learning through a range of supportive and challenging opportunities.

Pearson Science 2nd edition has been updated to fully address all strands of the new Australian Curriculum: Science which has been adopted throughout the nation. This edition also captures the coverage of Science curricula in states such as Victoria, which have tailored the Australian Curriculum slightly for their own particular students.

The *Pearson Science 2nd edition* features a more explicit coverage of the curriculum. The activities enable flexibility in the approach to teaching and learning. There are opportunities for extension as well as reinforcement of key concepts and knowledge. Students are also guided in self-reflection at the end of each topic.

More **explicit scaffolding** makes learning objectives clear and includes regular opportunities for reflection and self-evaluation.

In this edition, we provide a structured approach that integrates a seamless, intuitive and research-based approach hence **differentiating** the course for every student.

The Activity Book also provides richer application opportunities to take the Student Book content further with explicit coverage of Inquiry Skills, Science as a Human Endeavour and Science Understanding.

The diverse offering of worksheets allows students to be challenged at their level. Students have the flexibility to be self-paced and this new edition comes with the advantage of each worksheet being self-contained.

Be guided

A new handy **Toolkit** at the beginning of the Activity Book has been created to build skills in the key areas of practical investigations, research, thinking, organising, collecting and presenting. Each skill developed in the toolkit is directly relevant to applications in questions, investigations and research activities throughout the student and activity books. A toolkit spread provides guides and checklists alongside models and exemplars.

Be supported

Vocabulary boxes provide definitions for key terms, within the relevant context of the task. **Hints** help students get started on a worksheet and provide support in overcoming a barrier.

Be reflective

The **Thinking about my learning** feature provides the opportunity for self-reflection and self-assessment. It encourages students to look ahead to how they can continue to improve and assists in highlighting focus areas for skill and knowledge development.

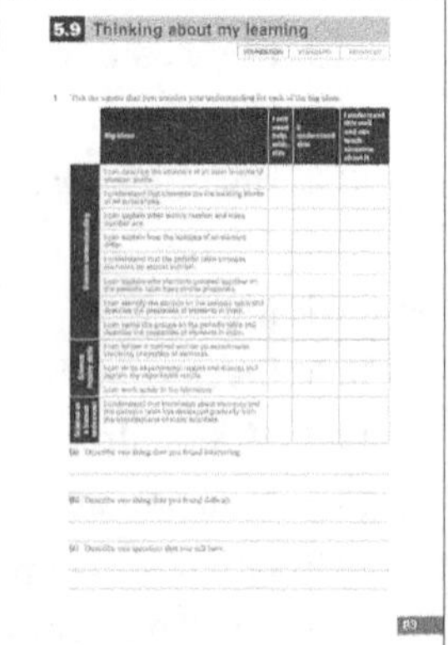

Be ready

A **knowledge preview** at the beginning of every chapter, activates prior knowledge relevant to the topic, providing an opportunity for students to show what they currently know. This handy tool supports teachers in assessing students' prior understanding.

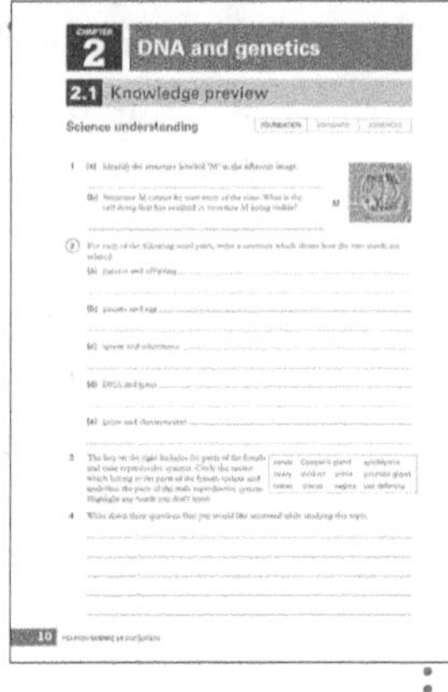

Be literate

Newly improved **literacy reviews**, in consultation with our Literacy Consultant Dr Trish Weekes, provide a deeper and broader range of language building tasks. Every chapter concludes with a literacy review which focuses on building a deeper understanding of key terms and supporting students to correctly apply key terms from the topic.

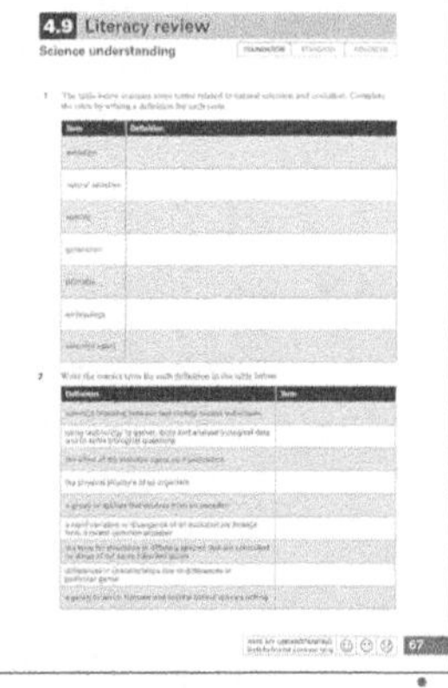

Be set

Visit www.pearsonplaces.com.au to enjoy the benefits of the following digital assets and interactive resources to support your learning and teaching:

- New interactive activities and lessons
- New Untamed Science videos
- Web destinations
- Student investigation templates and teacher support
- New STEP-UP student book and activity book chapters with answers at Years 9 & 10
- Full answers to all Student Book and Activity Book questions
- SPARKlabs
- Risk assessments
- Full teaching programs and curriculum mapping audits
- Chapter tests with answers

Each worksheet is classified according to the degree with which it deals with curriculum understandings.

- **Foundation** indicates the focus is on the basics like terminology.
- **Standard** indicates a focus on the core ideas, understandings and skills.
- **Advanced** indicates transfer and extension of core science understanding and skills to new or more sophisticated situations.

Teachers may use this tool to **differentiate** the worksheets allocated to students. They may select worksheets for students based on whether basic, core or extension exercises are required. The categories do not indicate the degree of difficulty of tasks on the worksheet. A worksheet labelled advanced may have tasks ranging from lower-level through to higher-level thinking.

Science understanding

FOUNDATION | STANDARD | ADVANCED

The main science **strand** is identified for each worksheet: Science Inquiry Skills, Science Understanding, Science as a Human Endeavour

Science Inquiry Skills relevant to the worksheet are identified: Questioning and predicting, Planning and conducting, Processing and analysing, Evaluating, Communicating

Each question in a worksheet is identified according to the degree of difficulty. Bloom's taxonomy is used as the basis for question classification. The classification is intentionally subtle and unobtrusive. This tool may be used to differentiate between students, matching questions to their levels of ability.

Question 1 is a straight number. This indicates a remembering or understanding lower-order question.

Question 5 has a circle around the question number. This indicates an analysing or applying middle-order question.

Question 7 has a square around the question number. This indicates a creating or evaluating high-order question.

1 Read through the words in the vocabular
(a) List the words that you know and wr

(5) Identify the solutes that did not dissolve i

[7] Discuss how this story illustrates that scie

An innovative tool for students to quickly and easily reflect on their understanding of each worksheet. The teacher may use the student responses as a formative assessment tool. At a glance teachers may assess which topics and which students need intervention for improvement.

RATE MY UNDERSTANDING
Shade the face that shows your rating

Science Toolkit

Scientific research

Sir Isaac Newton is probably among the most famous of all scientists. He made significant advances in physics, astronomy and mathematics. When praised for his discoveries he is quoted as writing 'If I have seen further, it is by standing on the shoulders of giants.'

This quote demonstrates the nature of scientific research and discovery. No scientist works in isolation. They work with others and build on the research done by their colleagues and their predecessors. Before scientists start experimenting, they investigate research already undertaken and review conclusions drawn in their area of interest. In this way, advances in science continue by building on what was learned before.

Understanding the research process

Research investigations involve a number of steps as shown in the Figure 0.1 flow chart. You can start to follow the steps in order but as the research project goes along, you can move between steps, not necessarily following them in the strict order of the chart. For example, if you realise that while checking your finished work you have not covered all aspects of the task, then you might need to find more resources, take more notes and write the new information into the report.

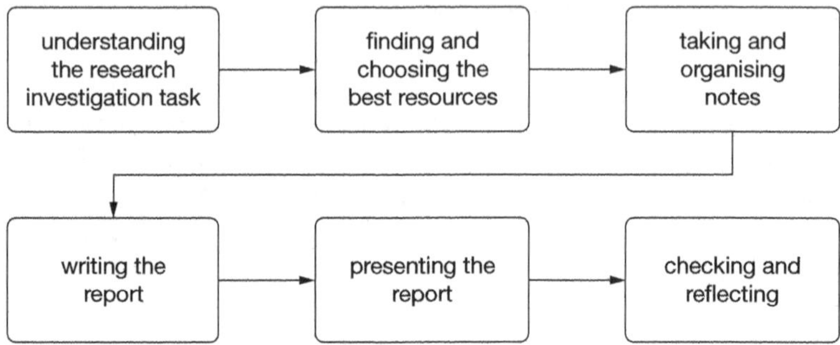

Figure 0.1 Research investigation steps

Understanding the research investigation task

It is essential that you fully understand what the research involves before commencing your investigation. Ask yourself the following questions:

Research investigation task questions to ask yourself		YES	NO
Research question or task	Do I understand the purpose of the research?		
	Do I understand the vocabulary in the set task or question?		
	Do I understand the research task or question?		
	Am I able to rephrase the research question/task?		
Time management	Do I know the due date of the science report?		
	Do I have a plan for managing my time that includes all stages of the research process?		

If you answered 'no' to any of the questions above, then the following actions will help you understand the research question, focus and get organised:

Research investigation task points to consider

- Read the task or question carefully.
- Look up the definitions of unknown words. Some key instructional words are listed in Table 0.1. Other vocabulary will need to be looked up.

Key word	Meaning
analyse	identify components and the relationships between them; draw out and relate implications
calculate	ascertain/determine from given facts, figures or information
classify	arrange or include in classes/categories
compare	show how things are similar or different
construct	make, build, put together items or arguments
contrast	show how things are different or opposite
define	state meaning and identify essential qualities
describe	provide characteristics and features
discuss	identify issues and provide points for and/or against
distinguish	recognise or note/indicate as being distinct or different from; to note differences between
evaluate	make a judgement based on criteria; determine the value of
explain	provide a sequence to make the relationships between things evident; provide why and/or how
justify	support using an argument or conclusion
list	write down phrases or items without further explanation
outline	sketch in general terms; indicate the main features of
predict	suggest what may happen based on available information
propose	put forward (for example a point of view, idea, argument, suggestion) for consideration or action

Table 0.1 Key instructional words

- Seek teacher assistance or discuss the topic with classmates to fully understand what is involved.
- Rephrase the question into your own words. It is very difficult to rephrase if you do not understand the meaning of the original. See the Table 0.2 example on page x.
- Break the question into smaller, manageable parts. This is especially important for complex tasks.

Research task example
To what extent is human activity responsible for climate change?
Terms that may need to be defined
extent: the size or the degree
responsible: to be given blame or credit for something
climate change: an alteration to world long-term climate
Rephrasing of the task
Climate change is caused by a variety of factors. Human activity can cause climate change. How much can climate change be blamed on human activity?
Smaller parts that task can be broken into:
• explain climate change • possible causes of climate change looking at different points of view • comparison of the different points of view • judgement of how responsible humans are for climate change, supported by evidence

Table 0.2 Breaking down and rephrasing tasks or questions example

- Draw up a timetable to plan how you propose to use your time to complete the task. A general guideline for time management is shown in Figure 0.1.

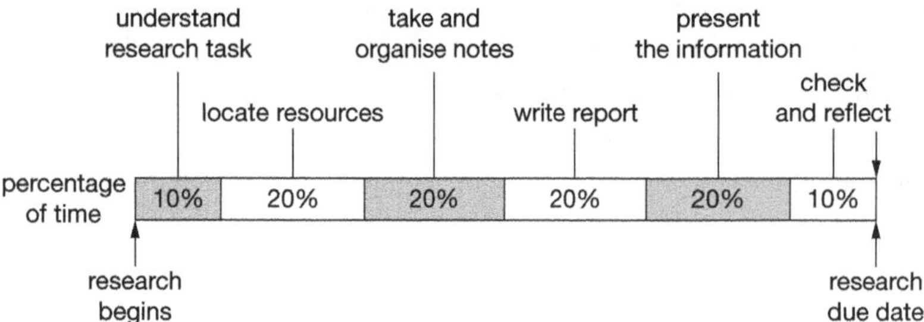

Figure 0.1 Time management guideline

Finding and choosing resources

Look carefully for the sources you may use in the research. Ask yourself the following questions:

Resource questions to ask yourself		YES	NO
Finding information	Does the research task require that I find secondary sources?		
	Does the research task require that I conduct an experiment?		
	Do I know where to look for resources?		
	Do I know how to conduct an experiment?		
	If I find one good resource, is that enough research?		
Choosing information	Do I know the most suitable types of resources I should look for to complete this research task?		
	Do I know how to evaluate websites to select the authoritative ones?		
	Should I rely entirely on internet resources?		
	Should I rely entirely on book resources?		
	Should I rely on a combination of book, internet and other resources?		

Resource points to consider

- Common places to look for resources are libraries, the internet, science text books, science magazines.
- Sources come in many forms such as print, the internet, expert speakers, television, documentary films, newspapers, magazines, surveys, experiments, interviews, samples. The type/s or resources you look for will depend on the nature of the research you are doing.
- Books and the internet are the most commonly used resources. Table 0.3 provides advantages and disadvantages in using these two common types of resources.

	Book resources	Internet resources
Advantages	– written by experts – authoritative information – proofread and accurate information – logical, organised layout – information accessible via title, contents page, index	– quick and easy to access – can access information which might otherwise be hard to find – accessible anywhere in the world – millions of websites provide information – can be quickly updated so may be the most recent information
Disadvantages	– a book takes some time to update and publish so check date of publication – accessibility may be limited because a book can only be used by one person at a time	– a search may lead you to a lot of irrelevant or biased websites – searches do not necessarily display the most useful websites – may be time consuming looking for information – fewer than 10% of websites are educational – can be difficult to tell if information is up-to-date, accurate or if it is written by an authority on the subject – information is not always ordered

Table 0.3 Advantages and disadvantages of book and internet resources

- No matter how good a resource is, it is wise to look more widely for resources, to gain different points of view and for different explanations of the science. What appears to be an excellent resource may be biased or contain incorrect information. It is good practice to compare resources and verify information.
- With so much information available, carefully evaluate which resources are the best for your research task. Do this by quickly scanning titles, subtitles, contents pages, readability and checking that diagrams and pictures are clear, well labeled and relevant.
- Carefully evaluate websites using the checklist in Table 0.4. Remember—anyone can publish anything so it is important to be internet literate.

Website evaluation checklist	
Credibility	Do I know who the author is?
	What are the author's qualifications or expertise?
	Does the author provide an email address or contact details?
	Does the web address indicate the site is reliable? Common web address endings are .edu (education site), .gov (government), .com (commercial).
Currency	What date was the site created?
	Is the information up-to-date?
Content	Is the information fact or opinion?
	Is information accurate? If you are unsure, check other resources to compare.
	Is there a bibliography?
	Are there spelling and grammar mistakes?
	Are there links to other quality sites?

Table 0.4 Website evaluation checklist

Taking and organising notes

Before you start to take and organise your notes, ask yourself the following questions:

Taking and organising notes: questions to ask yourself		YES	NO
Taking notes from resources	Do I know how to avoid plagiarism when I take notes?		
	Do I know what techniques I can use to take notes?		
	Am I sure that I can track where I got my notes from for a bibliography?		
	Are my notes concise and focussed on the task?		
Organising notes	Am I clear about how to organise my notes in a way that will make it easier to respond to all parts of the question?		

Taking and organising notes: points to consider

- Make sure you understand what plagiarism is so you know what to avoid. Table 0.5 shows an example of an original text, the same text plagiarised and an acceptable rephrasing of the text.

Original text	Plagiarised text	Rephrased text
Dogs have sweat glands on their feet. Dogs pant when they are hot because their sweat glands are not sufficient to cool down their bodies. In addition, their tongues allow the water from their bodies to evaporate and cool down their bodies.	Dogs have their sweat glands on their feet. When dogs get hot they pant because their sweat glands are not enough to cool down their bodies. Their tongues let the water from their bodies evaporate and cool their bodies.	Dogs pant in order to cool down. Water evaporates off their tongues and this lowers their body temperature. They cannot get cool enough from just the sweat glands on their feet.

Table 0.5 Example of original, plagiarised and rephrased text

- Do not copy or cut-and-paste slabs of information from sources. This is especially tempting from internet sources.
- Read small sections of the source at a time, then do not look at the source again as you take your summary notes.
- Take summary notes in your own words. Do not look back at the original sources when you use your summary notes to write up the research report.
- Common approaches to notetaking are summarised in Figure 0.2. The method you use is a personal preference. The example below responds to the question 'What are fungi and how do they grow?'

Original text

Fungi

Although fungi are technically microbes, many of them are large enough to be seen without a microscope. They often look like plants but do not use photosynthesis. Instead they feed on dead and decaying material, breaking it down further and helping chemical elements to return to the natural environment. Mushrooms, toadstools, yeasts and moulds are different types of fungi.

<table>
<tr><td colspan="1">Highlight/underline
– Read the original text
– Identify the key ideas about the topic
– Highlight or underline those key points</td><td colspan="2">Fungi
Although fungi are technically microbes, many of them are large enough to be seen without a microscope. They often look like plants but do not use photosynthesis. Instead they feed on dead and decaying material, breaking it down further and helping chemical elements to return to the natural environment. Mushrooms, toadstools, yeasts and moulds are different types of fungi.</td></tr>
<tr><td rowspan="4">Point form
– Read original text
– List headings under which information is needed
– Summarise information in dot points under listed headings</td><td colspan="2">Fungi</td></tr>
<tr><td>What fungi are</td><td>• microbes</td></tr>
<tr><td>Examples</td><td>• mushrooms
• toadstools
• yeast
• moulds</td></tr>
<tr><td>How they grow</td><td>• break down dead, decaying material
• don't use photosynthesis</td></tr>
<tr><td>Graphic organiser
Read original text
– Note key ideas in a visual form using an appropriate chart
– A variety of charts are available for use such as flow charts, concept maps, spider maps, Venn diagrams, fish bone map, tables</td><td colspan="2">Fungi 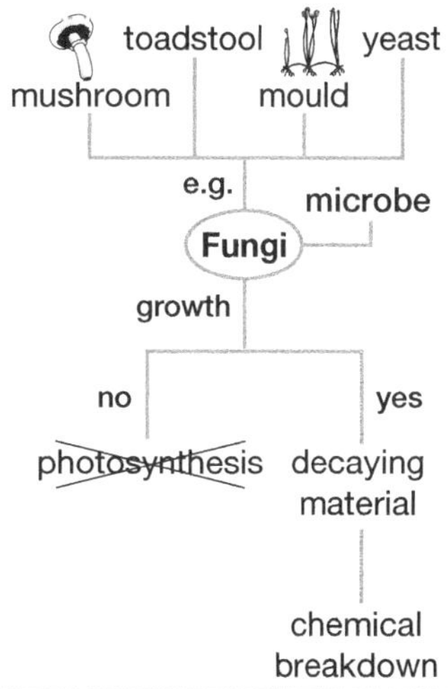
toadstool
yeast
mushroom
mould
e.g.
microbe
Fungi
growth
no
yes
photosynthesis
decaying material
chemical breakdown</td></tr>
</table>

Figure 0.2 Notetaking example

- Use a large A3 page grid to note-take when research tasks are complex. Notes can then be broken down into the sub-sections. Information from multiple sources can be collated and written onto the same grid, and bibliographic details can be noted. Table 0.6 on page xiv is a notetaking grid for the research task 'Contrast the roles that nitrogen-fixing bacteria and denitrifying bacteria play in the nitrogen cycle'.

Topic: Contrast the roles that nitrogen-fixing bacteria and denitrifying bacteria play in the nitrogen cycle			
	Source 1 (book) Title: Author: Publisher name: Publisher location: Date of publication:	Source 2 (website) Title: Author: Web address: Date accessed:	Source 3
Describe nitrogen cycle	research notes go here		
Explain what nitrogen-fixing bacteria are and what they do			
Explain what denitrifying bacteria are and what they do			
Differences between nitrogen-fixing bacteria and denitrifying bacteria			

Table 0.6 Notetaking grid

Writing the report

Before you commence writing your report, ask yourself the following questions:

Report writing questions to ask yourself	
Writing up the research task	How can I avoid plagiarism when I write up my report?
	How do I convert my research notes into a report? What is the correct language and writing style for a science report?
	How do I write up the bibliography?

Report writing points to consider

- Avoid referring back to the original sources so you are not tempted to copy the structure and expression of that text.
- Use formal language, expressed in the factual and passive voice. Do not use slang, conversation-style writing or emotional language. For example, it is acceptable to write 'Digestion is a complicated process beginning in the mouth with saliva breaking down carbohydrates.' It is not acceptable to write 'Your spit can't do all the work of digestion, but it starts things off in your mouth.'
- The research notes provide the foundation for the final report. Always reread the research task before beginning to write up the report.
- Research notes that have been carefully organised are more easily written up as a report. Organise the report into logical and sequential sections. Depending on the type of presentation format, diagrams and subheadings may be included in the body of the report or in an appendix at the end. Look at the simple example in Table 0.7.

<table>
<tr><th colspan="3">Research question: In what ways are sedimentary, igneous and metamorphic rocks similar and different?</th></tr>
<tr><td colspan="2">Research notes:</td><td rowspan="7">Final research report:
The Earth's three rock types are sedimentary, igneous and metamorphic. These rocks are all part of the rock cycle, a process where the material of the Earth's crust is very slowly created and destroyed through erosion and weathering. In this process sedimentary, igneous and metamorphic rocks are formed, broken down and new rocks remade. Rocks are hard and non-living objects.
There are many differences between the three rock types. Firstly, the way that these rocks form is different. While sedimentary rocks form when sediments settle and are compressed under water, igneous rocks form because of volcanic eruptions. Metamorphic rocks are different again, forming due to intense heat and pressure.
A second difference is where these three rock types form. Sedimentary rocks form on the Earth's surface beneath water. Particles that collect and are cemented together over long periods of time make layers of sedimentary rocks. Igneous rocks may form on the Earth's surface when a volcano erupts. These rocks may also form underground when pools of magma slowly cool. The heat and pressure needed to form metamorphic rocks means they form underground. It is sedimentary of igneous rocks that transform to metamorphic rocks.
A third difference . . . [Report continues, itemising each point of difference and explaining it.]</td></tr>
<tr><td colspan="2">Common features: sedimentary, igneous and metamorphic</td></tr>
<tr><td></td><td>• non-living
• part of rock cycle
• hard
• can be eroded/weathered</td></tr>
<tr><td colspan="2">Differences:</td></tr>
<tr><td>Sedimentary</td><td>• formed under water
• eroded particles settle and are compressed
• forms in layers
• may contain fossils
• particles from fine to pebbles
• mudstone, sandstone, conglomerate</td></tr>
<tr><td>Igneous</td><td>• formed by volcanic activity
• molten rock (magma) that solidifies (lava)
• form above and below ground
• larger crystals with slower cooling
• granite, pumice, obsidian, basalt</td></tr>
<tr><td>Metamorphic</td><td>• formed by heat, pressure beneath the Earth
• transformed sedimentary and igneous
• slate, marble, gneiss</td></tr>
</table>

Table 0.7 Report preparation and written example

- Acknowledge all resources you referred to for ideas and information in your research.
- The bibliography is placed at the end of the written report. There are a number of variations of suitable referencing techniques. Each type of bibliography has particular rules associated with how it is presented. The examples shown in Table 0.8 on page xvi demonstrate a method suitable for use in secondary school. Resources listed must be in alphabetical order. Remember to take particular care with punctuation.

	Books	Websites
Details needed	Author's surname and initials (Year of publication) *Title of book* City of publication State or country Publisher	Author OR organisation responsible for site (Year website was written or revised) *Title* Date website was retrieved URL
Example: 1 author	Smith, A. (2002). *How to Grow Plants.* Sydney, Australia: Pearson Australia	Pearson Australia. (2016). *Pearson Places.* Retrieved 1 August 2016, from http://www.pearsonplaces.com.au
Example: 2 or more authors	Laidler, G. et al. (2010). *Science and Inquiry* 2, Melbourne, Australia: Pearson Australia	
Example: author not clear	*Pearson Science 9 Student Book* (2001), Melbourne, Australia: Pearson Australia	

Table 0.8 Resources grid

Presenting the report

Before your present your report, ask yourself the following questions:

Report presentation questions to ask yourself	
Presentation format	What presentation formats do I have to choose from for writing up the research?
	What are the conventions for the way that format is normally done?

Report presentation points to consider

- A common way scientists present any new findings is by presentation of a poster at a scientific conference.
- The science poster gets ideas across to a large audience, in a limited space so it must be organised, succinct and creative. The information should be able to be read from at least a metre away and should not be crowded. A bibliography must be included and be visible. The checklist in Table 0.9 itemises the elements of an effective poster.

<table>
<tr><th colspan="2">Elements of an effective poster</th></tr>
<tr><td colspan="2">General
• suited to the audience
• large enough to be seen from 1 to 1.5 metres away
• correct spelling and good sentence structure
• if handwritten, then writing must be legible
• include a bibliography as an acknowledgement of all resources referred to in your research investigation
• title of a poster should be as short as possible
• has an introduction</td></tr>
<tr><td>Layout
• balance the display between visuals and text
• keep about 30% of poster blank with no text or visuals; this separates sections of information and helps the eye move around between information
• title positioned in centre top
• use subheadings
• keep visuals close to text they relate to
• organised to lead the eye through the display</td><td>Visuals
• only use relevant visuals that support your topic
• use quality visuals
• include captions for visuals
• all captions and other text need to be horizontal, not vertical
• simple and bold visuals; avoid unnecessary visuals</td></tr>
<tr><td>Font
• develop and display information in a hierarchy using font size and styles: key points in larger font; explanations in smaller font or dot points; main headings in a bigger sized font than subheadings; subheadings bigger font than other text
• do not use more than three different fonts on a poster</td><td>Colours
• black text on white background has better readability than on coloured backgrounds
• use only two or three related colours
• use neutral background colour
• restrict the use of bright colours that detract eye from the visuals and text; red, yellow and orange can be overpowering</td></tr>
</table>

Table 0.9 Effective poster checklist

- The checklist in Table 0.10 on pages xvii–xviii provides general information about other presentation formats. If the presentation includes an experiment, use the format for presenting the experiment results.

Presentation formats		
Format	**Characteristics**	**Things you should include or remember**
Pamphlet	• a small folded booklet with facts and diagrams • a summary of ideas • an easy way to access information • suitable for providing take-home information to individuals	• consistent layouts and fonts • dot points (if appropriate) • diagrams • attractive presentation
Report	• a presentation of clear and detailed information on a topic • suitable for providing detailed information to individuals	• an introduction, paragraphs and a conclusion • subheadings • mainly text but can include diagrams, maps, tables and/or graphs

Essay	• formal writing • can be used for discussion of a topic or to argue a point of view • suitable for formal writing to present to a single individual	• start with an introduction that gives the focus of the essay • continue with a series of paragraphs—each paragraph should make a new point and support it with evidence • start each paragraph with a topic sentence that states the point of the paragraph • end each paragraph with links back to the topic sentence and forward to the next paragraph • end the essay with a conclusion that ties all the ideas together but does not include any new information
Slide show	• an easy-to-follow format • a good format for presenting to a large audience • a useful tool if used with an oral presentation • able to provide a summary of information covered in an oral presentation • able to be printed as a set of notes to be given to the audience	• visual and written information • consistent backgrounds, formats and colours throughout • minimise the amount of text on any one slide
Oral presentation	• able to inform and possibly persuade the audience • engaging and can be entertaining • a good format for presenting a point of view on an issue • suitable for presentation to a large audience	• use relevant anecdotes • use cue cards to glance at if needed • rehearse your speech so you don't read the cue cards word for word • keep eye contact with the audience • speak clearly and at a volume that can be heard • don't fidget or wriggle around • stand up straight and look confident
Model	• a three-dimensional visual display • a good way to provide information with very little text • suitable for communicating information to a lot of people	• a title, labels and explanations
Website	• able to present visual and written information • accessible to a worldwide audience • easy to follow	• hyperlinks to related information • multimedia, such as video clips and audio, if appropriate • consistent backgrounds, formats and colours throughout • headings that stand out • list all of the site hyperlinks on the main page • the author's name and credentials, and the date of publication

Table 0.10 Presentation format checklist

Checking and reflecting

Checking and reflecting questions to ask yourself		YES	NO
Check over the research task	Have I answered the research question?		
	Have I finished all parts of the task?		
	Have I checked spelling, punctuation and grammar?		
	Is my name on the research report?		
Learning reflection	What did I have difficulty with? How can I improve this?		

Checking and reflecting points to consider

- Identify where you encountered problems doing the task. Focus on at least one problem and devise a strategy for improvement. You may find it helpful to discuss this with your teacher.
- Reread the whole research project checking for accuracy, readability, expression and clarity of meaning.
- If there is time, put the research report aside for a day or two before checking it.
- Check your work by reading it out aloud.
- Beware of computer spell checks. They may not successfully correct the work, as shown in Table 0.11. All mistakes are underlined.

Student's original work	Student's work after spell check	Corrected work
Globle warming is causing extreme whether. As temperatures rises the see level will rise and desserts will get alot biger.	Global warming is causing extreme whether. As temperatures rises the see level will rise and desserts will get a lot bigger.	Global warming is causing extreme weather. As temperatures rise the sea level will rise and deserts will get a lot bigger.

Table 0.11 Spell checking

CHAPTER 1

Science investigation skills

1.1 Knowledge preview

Science understanding

FOUNDATION	STANDARD	ADVANCED

Science research skills are essential in all branches of science—chemistry, physics, biology, psychology and geology.

1 **(a)** Consider the science research skills listed below. Place them in the table according to how much you feel you know about them. You may not have skills in every column.

creating a reference list	generating first-hand data	writing a bibliography	writing in passive voice
selecting resources	using in-text referencing	organising information	evaluating reliability of resources
avoiding plagiarism	managing a research task	locating second-hand data	presenting information using different formats
comparing sets of data	reading and understanding text and data	writing a research project	checking finished project

I am fairly sure I know how to do this	I have some idea how to do this	I do not know how to do this

(b) For each science research skill that you have written in the first column on the left, write a short sentence which shows what you know about using it.

1.2 Scientific questions and resources

Science understanding

FOUNDATION | STANDARD | ADVANCED

Knowledge of the key terms used in scientific questions is an important skill for understanding what is required. The ability to select quality resources is another useful skill.

1 What is the difference between 'describing' and 'explaining'?

____

____

____

(2) The table below lists some definitions and key words that may be used in a research question. Complete the gaps in the table by writing the correct meanings or key words in the blank spaces.

Key word	Meaning
	arrange or include in classes/categories
	show how things are similar or different
construct	
	show how things are different or opposite
	state meaning and identify essential qualities
propose	
	identify issues and provide points for and/or against
distinguish	
	make a judgement based on criteria; determine the value of

(3) What do the following three terms mean in relation to the quality of resources referred to for a research task?

Reliable: ______________________________

Reputable: ______________________________

Biased: ______________________________

(4) State two sources where you would expect to find reliable, reputable and unbiased scientific information:

1 ______________________________

2 ______________________________

[5] Explain why you would expect the resources listed in question 4 to be reliable, reputable and unbiased.

____

____

____

RATE MY UNDERSTANDING
Shade the face that shows your rating   

1.3 Referencing sources of information

Science understanding

FOUNDATION | **STANDARD** | ADVANCED

Referencing acknowledges the work of other authors and their contribution to your work.

1. What is the difference between a reference list and a bibliography?

2. The following is a reference list that was used at the end of a research project. Rewrite this information below as a correctly presented reference list.

Calcium carbonate and belching, 2015, retrieved may 17, 2016 from http://www.livestrong.com/article/535708-calcium-carbonate-and-belching/.

Indigestion, Pain, Symptoms, remedies, causes and more. (no date). Retrieved May 17th, 2016 from http://www.webmd.com/heartburn-gerd/indigestion.

'Neutralization', 2013, retrieved May 17th 2016, from http://chemwiki.ucdavis.edu/Core/Physical_Chemistry/Acids_and_Bases/Acid%2F%2FBase_Reactions/Neutralization.

1.3 Referencing sources of information

(3) The following extract is from a student assignment.

Antacids are weak bases that consist of hydroxide and carbonate. They help to neutralise stomach acid, which can cause symptoms such as inflammation of the oesophagus. http://www.webmd.com/heartburn-gerd/indigestion.

'When an acid and base react together they produce salt and water.' Neutralization.

For example HCl + NaOH -> H_2O + NaCl. During the reaction, hydrogen ions (H^+) from the acid and (OH^-) ions from the base react together to form a water molecule. Salt is also formed when the anion from the acid and the cation from the base join together. Neutralization, 2013, http://chemwiki.ucdavis.edu/Core/Physical_Chemistry/Acids_and_Bases/Acid%2F%2FBase_Reactions/Neutralization.

(a) Identify and highlight the errors in the referencing technique.

(b) Rewrite the paragraph below using correct in-text referencing.

RATE MY UNDERSTANDING
Shade the face that shows your rating

1.4 Acknowledging the work of others

Science inquiry skills

FOUNDATION	STANDARD	ADVANCED

Communicating

Imagine you are researching one of the following topics:

- the existence of planets around stars other than our Sun
- animal testing by pharmaceutical companies
- whether vaccination of children should be compulsory.

1. Locate six resources you might use for the research task. Include at least two different types of resources (e.g. websites, books or magazines). Provide the bibliographic details of each resource.

2. Evaluate each resource you listed in question 1 by completing the table below. Provide a final judgement about whether you would or would not use this resource.

Resource title	Type of resource	Positive features of this resource	Problems with using this resource	Use/do not use this resource

RATE MY UNDERSTANDING
Shade the face that shows your rating  

1.5 Working with data

Science inquiry skills

FOUNDATION | **STANDARD** | ADVANCED

Processing & Analysing

Table 1.5.1 is taken from the Australian Bureau of Statistics and outlines the projected population growth for each Australian state that is expected by 2061.

Table 1.5.1 State-by state population statistics and projections

State	2012	% of national	2061	% of national	Average annual growth (%)
New South Wales	7305882	32.2	11475527	27.6	0.9
Victoria	5630855	24.8	10305516	24.8	1.2
Queensland	4568414	20.1	9259341	22.4	1.5
South Australia	1656454	7.3	2308149	5.6	0.7
Western Australia	2434738	10.7	6402253	15.4	2.0
Tasmania	512199	2.3	565710	1.4	0.2
Northern Territory	235233	1.0	435024	1.1	1.3
Australian Capital Territory	375076	1.7	740903	1.8	1.4

Source: ABS, RP Data

1 Is this data considered to be first- or second-hand data?

2 Can the source be considered reliable? Why or why not?

3 Can the predicted data for 2061 be considered to be precise?

4 Identify the state that had the largest population in 2012.

5 Identify which state is expected to experience the fastest annual percentage population growth.

6. On the blank graph, draw a column graph of the data for the total population of Australia for the years 2012 and 2061.

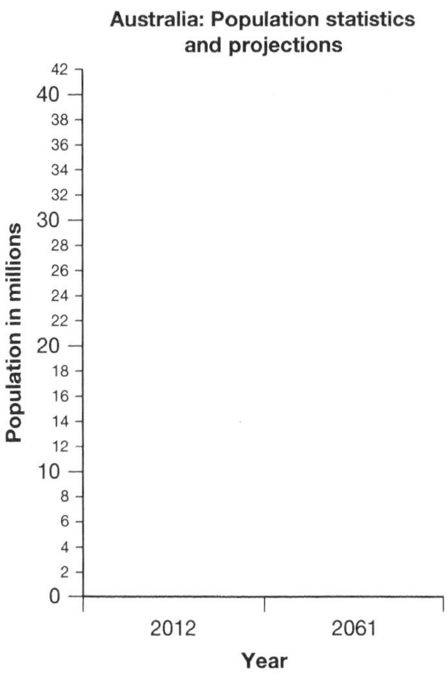

7. Calculate the change in Australia's total population between 2012 and 2061.

8. State one conclusion that could be made from this data about the distribution of people across Australia in 2012.

9. State one conclusion that could be made from this data about the distribution of people across Australia in 2061.

10. How is the population distribution across Australia predicted to change between 2012 and 2061?

RATE MY UNDERSTANDING
Shade the face that shows your rating

1.6 Literacy Review

Science understanding

FOUNDATION | STANDARD | ADVANCED

1 Read the clues and write the word that matches the clue.

Clue	Word
The branch of science that is concerned with living things.	
The branch of science concerned with matter and energy.	
The branch of science concerned with substances, their properties and reactions.	
The branch of science concerned with the human mind and its functions.	
The branch of science concerned with the physical structure of the Earth and the processes that work on it.	
The name of the type of data that is collected from scientific investigations.	
The name of the type of data collected from science text books.	
The practice of taking someone else's work or ideas and passing them off as your own.	
A brief statement of the main points of an investigation.	
A stated source of information.	
The following statement is the ___________ in a scientific investigation report: 'If the temperature of a cup of water is increased then more sugar will dissolve in it.'	
The list of materials in a scientific investigation is written just before the ______________.	
A list of all resources used in scientific research.	
Another term for numerical data.	
The way something is spread or shared out over an area.	
The description of the steps in a scientific experiment.	
The voice in which scientific investigation reports should be written.	

RATE MY UNDERSTANDING
Shade the face that shows your rating

1.7 Thinking about my learning

Think carefully about what you have learned about science skills and about how you approached this unit of work. Then complete the following sentences.

While learning about science skills I was surprised to find out that ...
The most interesting thing I learned was ...
One concept I found very easy to understand was ...
One idea I found quite challenging was ...
The strategy I used to help me to understand this difficult concept was ...
I am still not confident that I fully understand ...
In order to help me understand this better I am going to ...
While learning about science skills, the activity I enjoyed the most was ...
The activity I enjoyed the least was ...
When studying the next topic some things I will do differently are ...

CHAPTER 2

DNA and genetics

2.1 Knowledge preview

Science understanding

FOUNDATION | STANDARD | ADVANCED

1 **(a)** Identify the structure labelled 'M' in the adjacent image.

____

(b) Structure M cannot be seen most of the time. What is the cell doing that has resulted in structure M being visible?

M

____

(2) For each of the following word pairs, write a sentence which shows how the two words are related:

(a) parents and offspring ______________________________

(b) gamete and egg ______________________________

(c) sperm and inheritance ______________________________

(d) DNA and genes ______________________________

(e) genes and chromosomes ______________________________

3 The box on the right includes the parts of the female and male reproductive systems. Circle the names which belong to the parts of the female system and underline the parts of the male reproductive system. Highlight any words you don't know.

cervix	Cowper's gland		epididymis
ovary	oviduct	penis	prostate gland
testes	uterus	vagina	vas deferens

4 Write down three questions that you would like answered while studying this topic.

____

____

____

____

____

2.2 Structure of DNA

Science understanding

FOUNDATION | **STANDARD** | ADVANCED

DNA building blocks

The building blocks of a DNA molecule are called nucleotides. Each nucleotide has three parts:

- phosphate group
- sugar
- nitrogen-rich bases.

1 Use the three terms above to label the nucleotide diagram shown on the right.

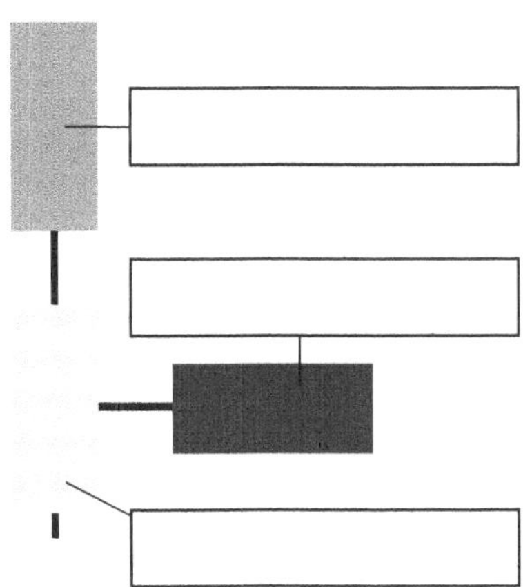

DNA structure

The nitrogen-rich bases can be one of four types:

- adenine (A)
- thymine (T)
- cytosine (C)
- guanine (G)

In a double-stranded DNA molecule, A and T always form a pair and C and G always form a pair. They are called complementary pairs.

complementary (*adj*) matching, one of a pair

DNA (*n*) deoxyribonucleic acid—main part of chromosomes; transfers genetic characteristics

double-stranded (*adj*) with two strands

strand (*n*) rope or thread

2 The sequence of bases below represents a section of a single strand of DNA. Write the base sequence in the complementary strand of DNA.

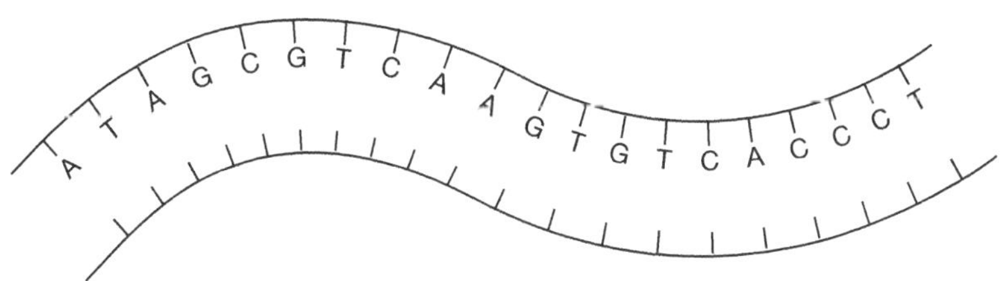

3 **(a)** Identify which of the following figures A, B, C, or D represents a possible correct base sequence in a molecule of DNA. ______________________

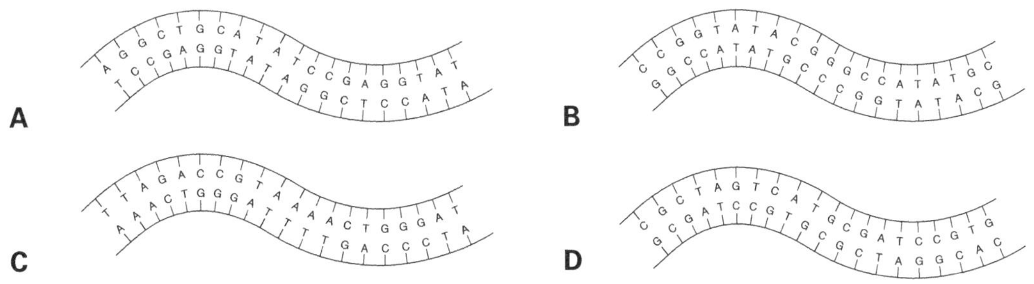

(b) Justify your decision for each molecule, explaining why it is correct or incorrect.

A ______________________

B ______________________

C ______________________

D ______________________

RATE MY UNDERSTANDING
Shade the face that shows your rating

2.3 Complementary base pairing

Science understanding

FOUNDATION | STANDARD | ADVANCED

The chemical structure of the nitrogen-rich bases means that they can only form chemical bonds with one of the other bases.

- Adenine (A) only pairs with thymine (T).
- Cytosine (C) only pairs with guanine (G).

One side of the DNA 'ladder' could be like Figure 2.3.1 with the sugar–phosphate backbone and the attached bases.

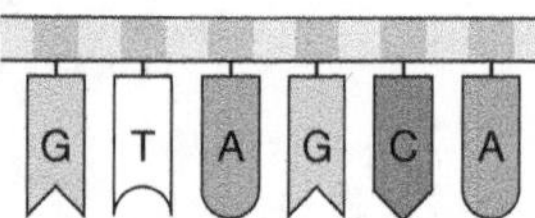

Figure 2.3.1 One side of the DNA strand

Using complementary base pairing, the other side of the molecule would look like Figure 2.3.2.

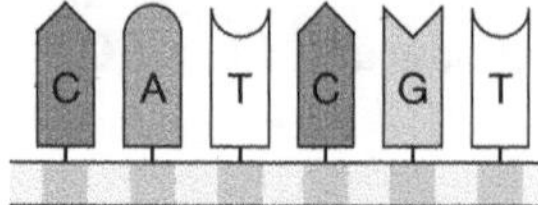

Figure 2.3.2 The other side of the DNA ladder

When the two sides are put together, the DNA molecule in Figure 2.3.3 would result.

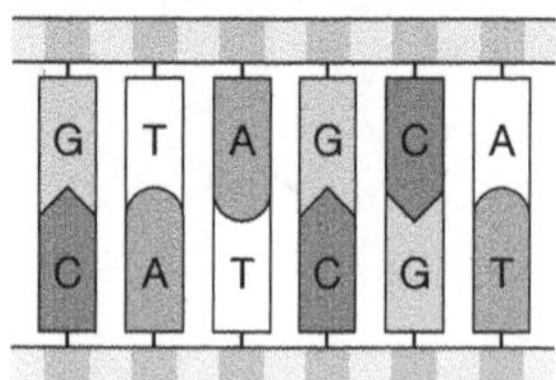

Figure 2.3.3 The resulting DNA molecule

(1) Construct a diagram of the complementary DNA ladder for the following sequences.

(a)

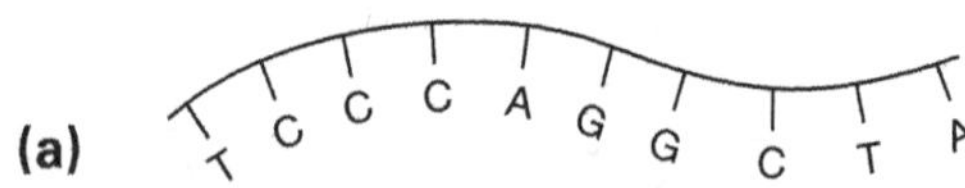

(b)

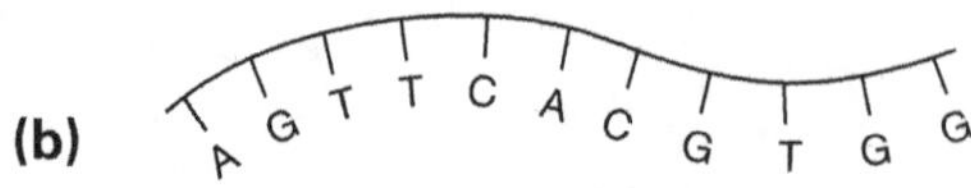

(c)

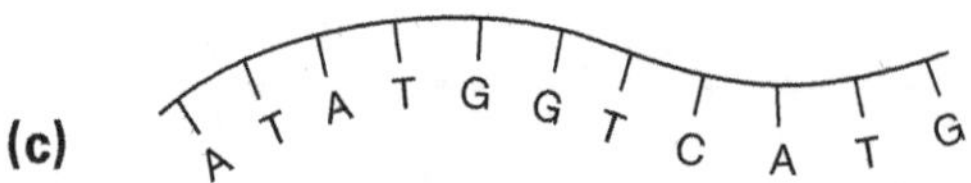

RATE MY UNDERSTANDING
Shade the face that shows your rating

2.4 Mitosis

Science understanding

FOUNDATION	**STANDARD**	ADVANCED

The diagram represents five stages of mitosis. However, they are not in the correct order.

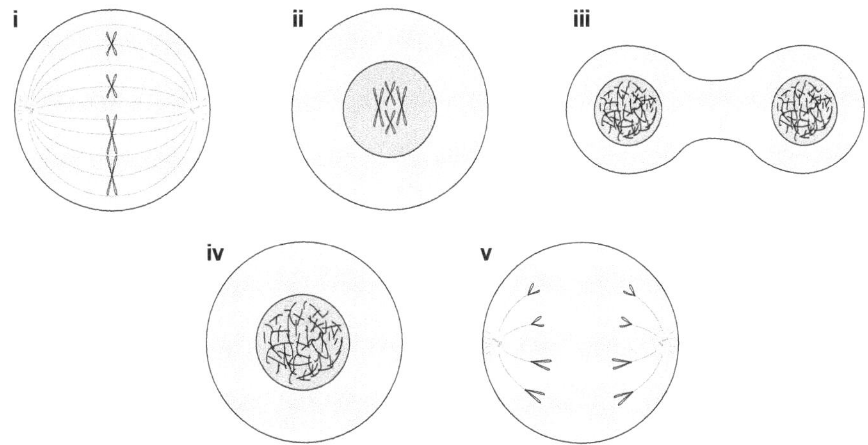

The following captions (A to E) are descriptions of the five stages of mitosis shown above. The descriptions are not in the correct order.

A The chromatids separate. The spindle fibres contract, pulling the chromosomes to opposite poles of the cell.

B Separate chromosomes become visible. Each chromosome comprises two chromatids.

C In the period between cell divisions, the DNA replicates.

D The nuclear membrane re-forms, enclosing the chromosomes into a new nucleus at each pole. Division of the nucleus is now complete. The cytoplasm then divides, resulting in two identical daughter cells.

E The membrane surrounding the nucleus breaks down. The spindle appears, extending from the poles of the cell to each chromosome. The chromosomes line up across the equator of the cell.

comprise (*v*) made up of, to consist of
contract (*v*) to shorten
divide (*v*) to split in half
equator (*n*) middle, centre
identical (*adj*) exactly the same
replicate (*v*) to reproduce or copy itself

1 Use the table below to redraw each stage of mitosis in the correct order.

2 Identify the letter of the correct caption for each stage.

First stage	Second stage	Third stage	Fourth stage	Fifth stage
Caption:	Caption:	Caption:	Caption:	Caption:

2.5 Meiosis

Science understanding

FOUNDATION	STANDARD	ADVANCED

1 Meiosis takes place in 8 different stages. Draw lines to match each diagram of meiosis on the right with its correct description on the left.

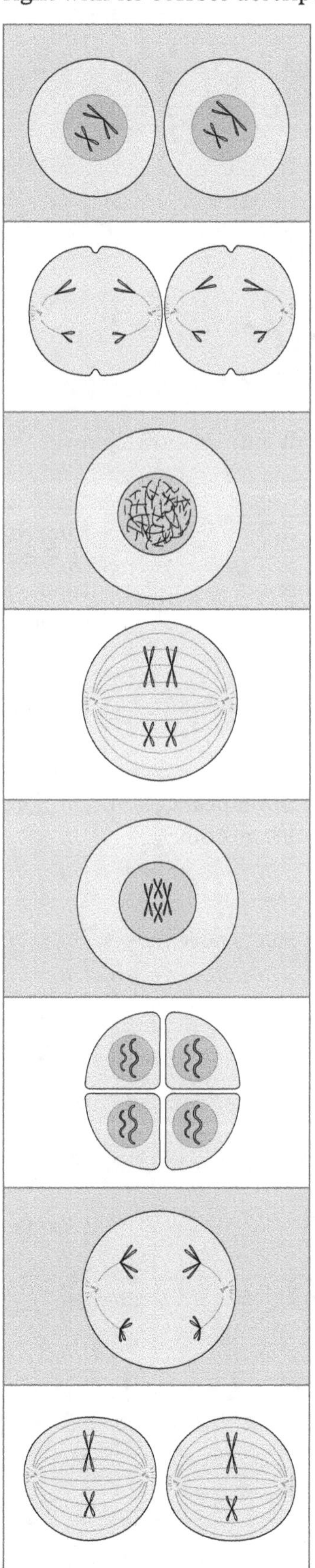

The nuclear membrane breaks down and pairs of replicated chromosomes line up on the equator of the cell with spindle fibres attached.

The nuclear membranes form and the cytoplasm divides to produce four new cells. Each cell now contains the haploid number of chromosomes. These cells are the gametes.

In the period between cell divisions, the DNA replicates.

The spindle fibres contract, pulling the chromatids apart towards the poles of the cells.

At the end of the first part of the division, nuclear membranes may re-form. The chromosomes at each pole are enclosed into new nuclei.

The spindle fibres contract, drawing one chromosome from each pair to opposite poles of the cell. At this stage, each chromosome is still two chromatids.

The DNA becomes visible as separate chromosomes, each of which comprises two chromatids.

After the first part of the division, a new spindle forms at right angles to the first. The new spindle fibres attach to the chromosomes and line them up on the equator of the cell.

RATE MY UNDERSTANDING
Shade the face that shows your rating   

2.6 Punnett squares

Science understanding

FOUNDATION | **STANDARD** | ADVANCED

You may have noticed that some people have long eyelashes and others have short, straight eye lashes. Having long or short lashes is the phenotype of the person—the way they look. The length of your eyelashes is an inherited trait, with long lashes being dominant to short lashes. The alleles that you have inherited to determine the length of your eyelashes is your genotype.

cross (*n*) the mating of two organisms
inherited (*v*) received from another individual such as a parent
offspring (*n*) child
trait (*n*) quality, characteristic

1. Use a Punnett square to demonstrate the inheritance of long and short eyelashes respectively.

   In this example, Ria, the mother, is homozygous for long lashes. Aidan, the father, is homozygous for short lashes.

   Use the letter E to represent the allele for long eyelashes and e to represent the allele for short eyelashes.

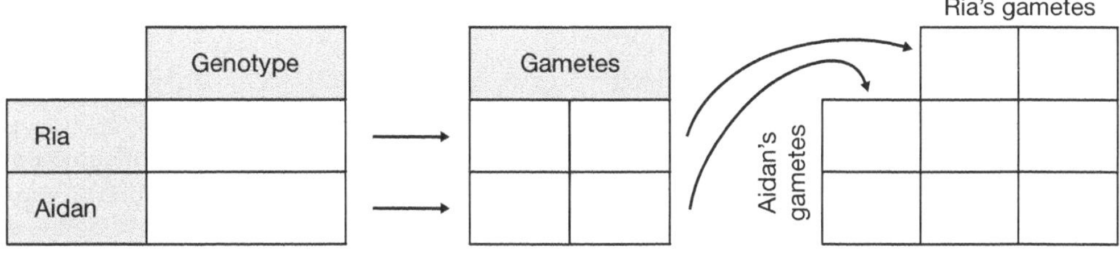

   (a) Possible genotypes of children: ______

   (b) Possible phenotypes of children: ______

2. Demonstrate how the characteristics of the offspring would change if both parents were heterozygous for eyelash length.

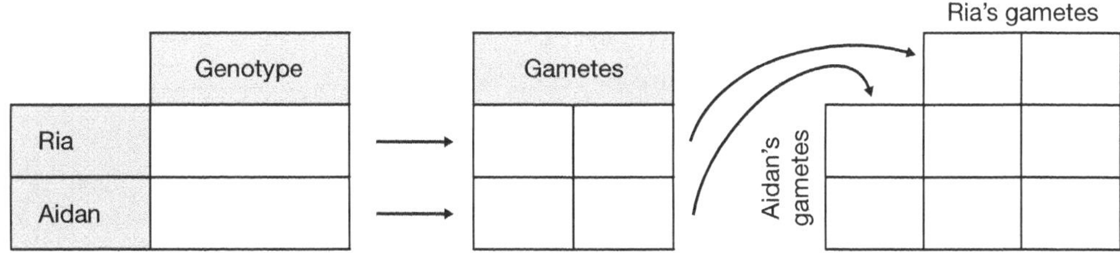

   (a) Possible genotypes of children: ______

   (b) Possible phenotypes of children: ______

3. In guinea pigs, black coat colour (B) is dominant to white coat colour (b).

   (a) What is the ratio of phenotypes and genotypes of the offspring from a cross between a heterozygous black guinea pig and a homozygous white guinea pig?

   ______

   (b) How did you work out the answer? ______

RATE MY UNDERSTANDING
Shade the face that shows your rating

2.7 Pedigree analysis

Science inquiry skills

FOUNDATION	STANDARD	ADVANCED

Processing & Analysing **Communicating**

Inheritance of a characteristic in a family can be demonstrated using a family tree or pedigree.

In a pedigree, such as the one shown, symbols are used to identify males and females, and those with or without a characteristic or trait.

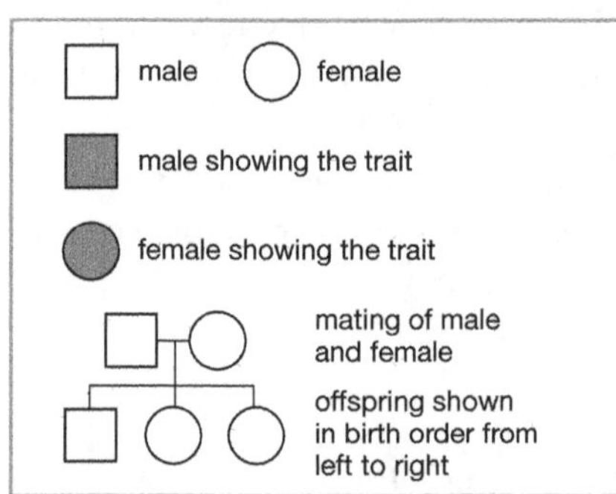

1. The following diagram shows a pedigree for three generations of a family in which the ability to roll the tongue has been recorded. Tongue rolling is a dominant trait.

   Determine the genotype of each individual in the pedigree and record it in the box provided. Use the letter R to represent the dominant trait. Use a '?' for an unknown allele (for example 'R?').

inheritance (*n*) a characteristic passed on from parent to offspring

generation (*n*) group of individuals born and living about the same time

pedigree (*n*) a chart showing parents, offspring and their inherited traits

roll the tongue (*v*) to curl up the sides of the tongue

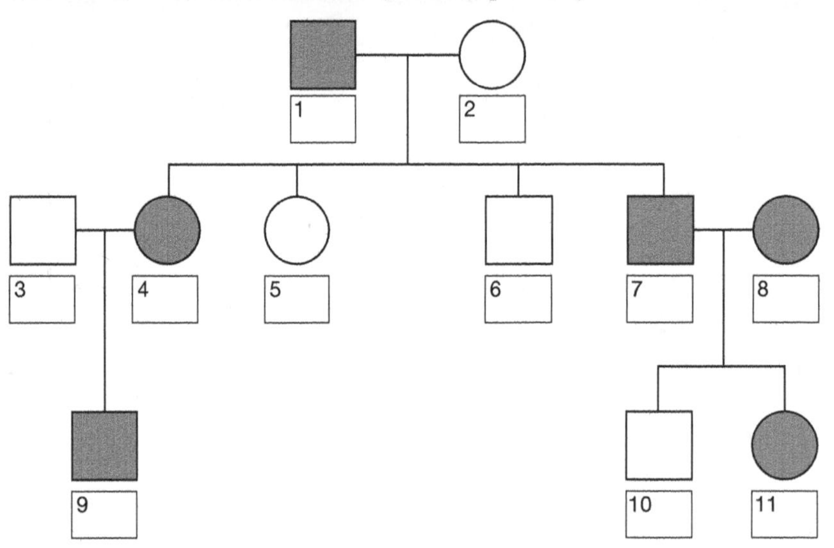

2. **(a)** The pedigree below shows inheritance of a different trait over three generations.

   Determine whether it is a dominant or recessive trait. ______________________

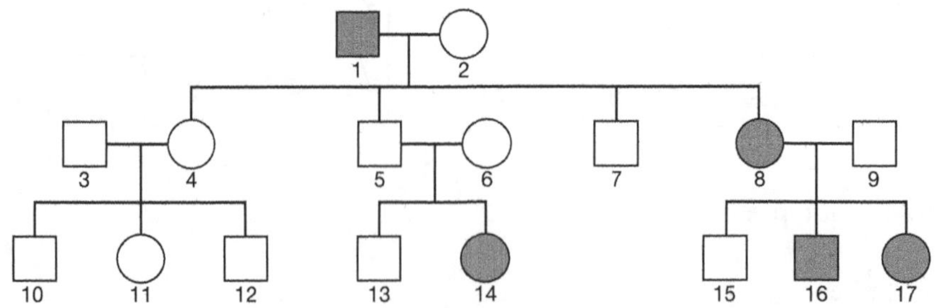

   (b) Justify your response.

   ____

   ____

   ____

   ____

3 The following family pedigree is for the recessive disease cystic fibrosis.

cystic fibrosis (*n*) an inherited disease that affects the lungs, pancreas and intestines

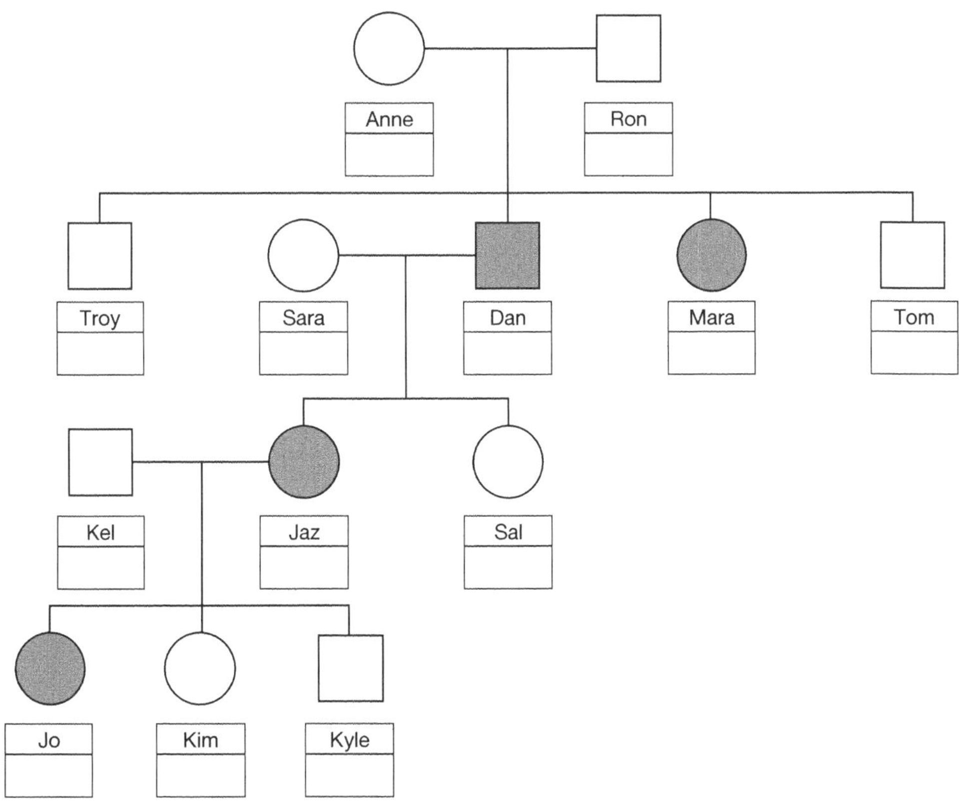

(a) Determine the genotype of each of the family members and record it in the space provided. Use the letters C and c to represent the alleles in the genotypes.

(b) Identify by name the two individuals for whom you could not work out the genotype.

____

(c) Explain why you were not able to work out the genotypes for these individuals.

____

____

____

____

RATE MY UNDERSTANDING
Shade the face that shows your rating

2.8 Sex-linked genes

Science inquiry skills

FOUNDATION | **STANDARD** | ADVANCED

Processing & Analysing | Communicating

The gene for coat colour in cats is carried on the X-chromosome. There are two alleles—black (B) and orange (O). These two alleles are codominant. This means that the characteristics of black and orange will both appear in the phenotype of an individual that is heterozygous.

The genotypes X_BX_B and X_BY result in black cats.

The genotypes X_OX_O and X_OY result in orange cats.

The genotype X_BX_O results in a tortoiseshell cat that has black, orange and white patches of fur.

1. Use a Punnett square to determine the possible genotypes of the offspring resulting from a cross between a black male and an orange female.

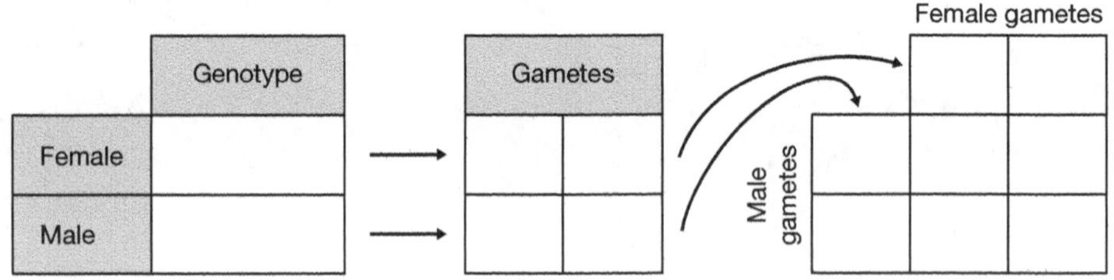

2. One of the tortoiseshell cats from the cross in question 1 had six kittens—one black female, three tortoiseshell females, one black male and one orange male.

   (a) Deduce the genotype of the father of the litter. ________________________

   (b) Use a Punnett square to justify your answer. ________________________

3. Construct a pedigree diagram showing the three generations of cats. Use the following symbols when drawing the pedigree.

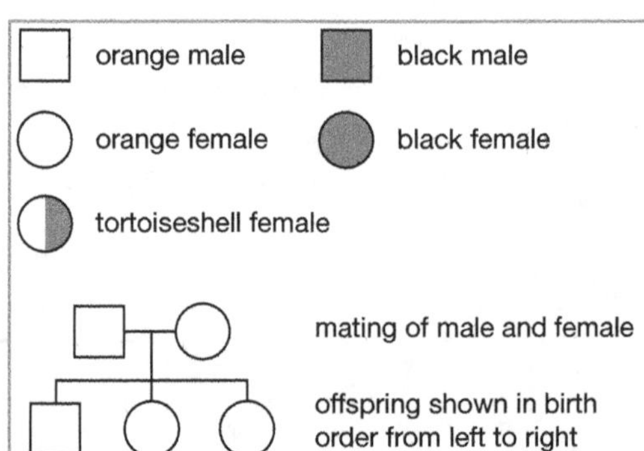

RATE MY UNDERSTANDING
Shade the face that shows your rating

2.9 Mutations

Science understanding

FOUNDATION | STANDARD | ADVANCED

Mistakes can occur as DNA is copied causing a change in the order of the bases along the DNA strand. This type of change is called a mutation. Some mutations happen naturally. Others are the result of damage caused by, for example, chemicals or radiation. Factors that have the potential to cause mutations are known as mutagens.

The effect of mutagens has been known for a very long time. More than 2500 years ago, long before chromosomes or DNA were discovered, a Greek philosopher called Hippocrates identified tumours in patients. He described these as cancer because the shape of the tumour resembled a crab.

Ultraviolet (UV) light from the Sun is a natural mutagen, as are some viruses. Tobacco smoke contains mutagens. Mutagens are also found in some foods and drinks.

Mutagens affect DNA in different ways. Some mutagens change a single base in the DNA code. These are called point mutations. An example is shown in Figure 2.9.1. Changing one base for another sometimes has little effect on the functioning of the gene.

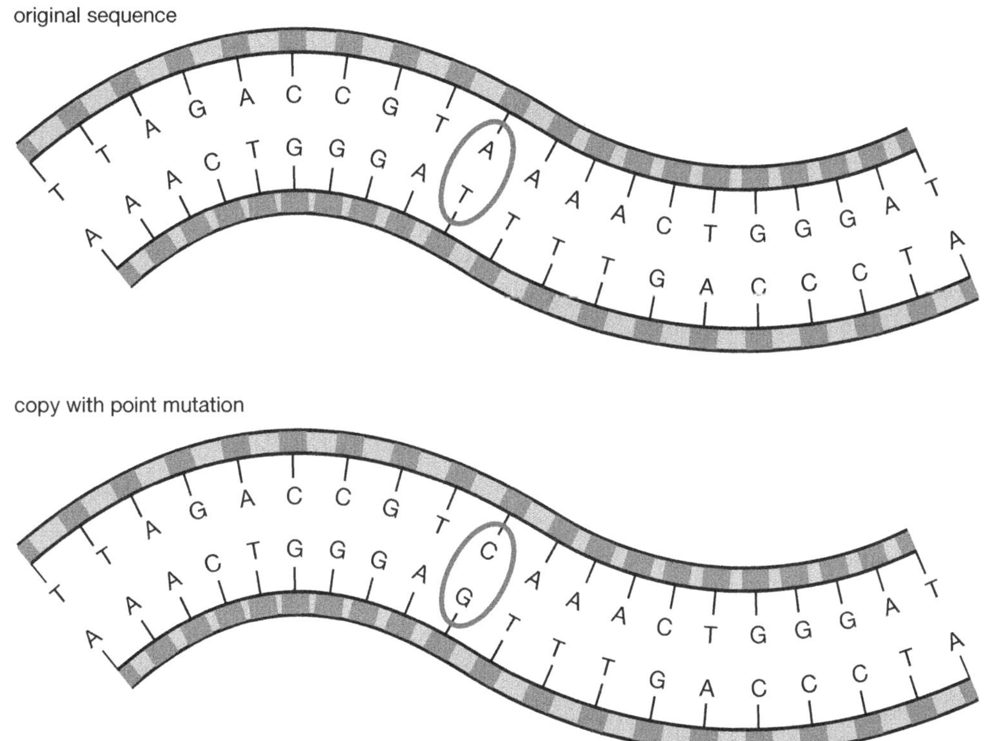

Figure 2.9.1 Example of a point mutation. In the original sequence, base A is paired with base T. When the DNA was copied, a single base in the DNA sequence was altered at that point. There is now a base C, which will pair with a base G.

Other mutagens delete or insert one or more bases into the DNA molecule causing a change in the sequence of bases that follow (see Figure 2.9.2). These are major changes to the DNA and are known as frameshift mutations.

2.9 Mutations

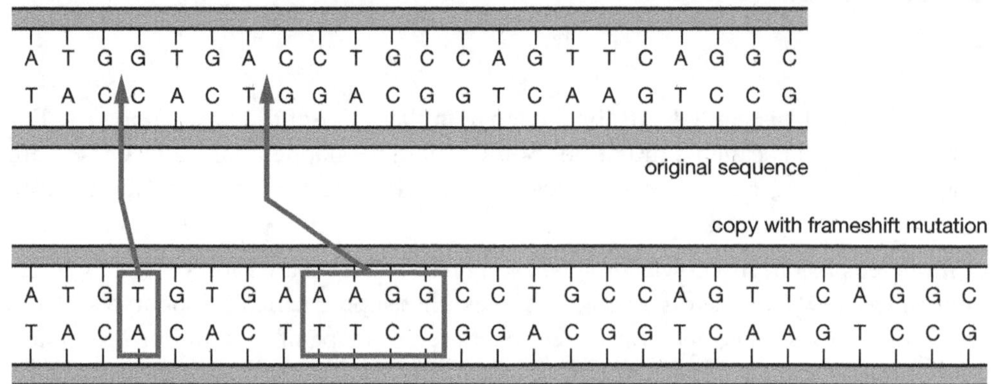

Figure 2.9.2 Example of a frameshift mutation. Two mutations have occurred in copying the original sequence. The first mutation shows where a new base pair (T-A) has been inserted into the original sequence; the second shows a major change where a whole new sequence has been inserted (A-T, A-T, G-C, G-C).

Infrequent, small doses of mutagens may have little effect. Many potential mutations are repaired before they can cause permanent damage. Proteins within the cell nucleus 'cut out' damaged sections of DNA and fill the gap with the correct sequence of bases. When mutations occur at a faster rate than the repair process can take place, some areas of damage may go uncorrected. The genetic information may be permanently mutated and disease may be the outcome. This can occur if the person is exposed to a large amount of the mutagen or exposure is continuous over a long period of time. For example, nicotine in cigarette smoke is thought to block the process of repairing damaged DNA in lung cells. The result of long-term cigarette smoking is lung damage and, potentially, lung cancer.

Normal cells respond to changes in their environment in particular ways. Cells with DNA that contains a mutation do not behave normally.

Mutations can occur in cells of the body (somatic cells) and they occur in the cells that produce gametes (germline cells). Mutations in somatic cells such as cells of the liver or bone marrow may damage the cell and change the functioning of the liver or bone marrow. However, they do not affect any offspring of that person.

Germline mutations are passed on to the next generation if the egg or sperm involved in fertilisation has the mutation. The mutation will be found in every cell of the new individual, including any gametes they produce. Germline mutations are passed from generation to generation.

1. Compare:

   (a) a mutation and a mutagen

   ____

   ____

   (b) a point mutation and a frameshift mutation

   ____

   ____

   (c) a somatic mutation and a germline mutation.

   ____

   ____

2. Explain how mutations occur without a mutagen being present.

3. Explain why it is impossible to avoid mutagens.

4. **(a)** Describe what normally happens in the body when a mutation occurs.

   (b) Explain what is different in the mutation process when a person is exposed to high levels of a mutagen.

5. Explain why mutagens that affect the ovaries or testes are of more concern than other types of mutagens.

6. When the genetic information is 'read' from DNA, each group of three bases is like a 'word'. The more 'words' that are changed the more significant the mutation. Changing one 'word' in the middle is a point mutation. Changing many words is a frameshift mutation. In the table below, complete the following tasks:
   - Classify these mutations as point mutations or frameshift mutations.
   - Identify the point at which the mutation occurred by drawing a circle around it in the second column.
   - Identify the number of 'words' that are changed by underlining those that are different in the mutated code.

	Single strand of original DNA code	Mutated code	Mutation type
(a)	AAT CGC GAT GGC TTA CCG	AAT CGC CAT GGC TTA CCG	
(b)	TTC CGA GGT CAG GAT CCA	TTC GCG AGG TCA GGA TCC	
(c)	AGG CAC GTT GCA AAT CGG	AGG CAC GTT GCA AAA CGG	
(d)	GGC CTA TGT GAA CCA GGT	GGC CTA TGA GAA CCA GGT	
(e)	GGC GGT TTA TGC CGA GCT	GGC GGT TTA ATG CCG AGC	
(f)	GCG ATC GGG ATA TCC ATA	GGG ATC GGG ATA TCC ATA	

2.10 Genetically modified food

Science understanding

FOUNDATION | **STANDARD** | ADVANCED

The Green Revolution of the 1950s increased food production by using new and improved chemicals to control weeds, insect pests and diseases. New varieties of crops and fertilisers also helped to increase food production.

The Gene Revolution of the 21st century uses genetic modification to grow crops that have the potential to produce more food with a higher nutritional value than traditional crops. The Gene Revolution also uses fewer chemicals. Scientists believe that by using gene technology they can improve a variety of crops including corn, wheat, rice, canola, chicory, squash, potato, soybean, alfalfa, cotton, banana and tomato.

(1) Compare the Green Revolution and the Gene Revolution.

____

____

Below are some of the arguments for and against the use of genetically modified (GM) food.

Arguments for the use of genetically modified crops

- GM crops are potentially more resistant to disease, can grow in less space, can provide greater yield and need less pesticide.
- Genetically modified, pest-resistant crops need less insecticide spray, which is better for the environment and saves the farmer money.
- Current agricultural methods will not be able to grow enough food to feed the 9 billion people predicted to populate the world by 2050.
- Genetic modification can improve crops more quickly than conventional selective breeding processes.
- Scientists can add genes that make plants tolerant to frost, drought and salinity (high salt levels). They can also add genes to make crops resistant to insect pests. These genes can be turned 'off' and 'on' in different parts of the plant. Genetic modification is one tool that farmers can use to maintain or increase crop yields as the climate changes.
- GM foods can improve a poor diet by providing nutritionally improved foods.
- Improved nutrition should have health benefits in both developed and developing countries. GM plants can also deliver medicines. For example, golden rice increases the intake of vitamin A, and bananas can carry a vaccine (cure) for the disease hepatitis D.
- Genetic modification may be able to remove allergens from nuts. Allergens are the chemicals that cause allergic reactions. Eleven different proteins in peanuts are known to cause allergic reactions. Scientists are developing genetically modified peanuts in which the two strongest allergens have been removed.
- There is no evidence to suggest that approved GM foods are more dangerous than normal foods.
- In Australia, GM foods are regulated, ensuring that only assessed and approved GM foods enter the food supply.

Arguments against the use of genetically modified crops

- Some people say that GM crops are not safe to eat. They feel that there has not been enough evaluation of the potential risks and side effects of consuming GM foods, especially those that are nutritionally boosted. There is a chance that new allergens may be created.
- Herbicides are chemicals that are used to control weeds. The genes for herbicide resistance may be transferred from the GM crop to weeds in the environment, making it more difficult to control weed species.

- Some people think that antibiotic resistance may develop in humans and farm animals fed on genetically modified foods. This could make antibiotics less effective in treating disease.
- Creating pest- or herbicide-resistant GM crops could lead to the evolution of more-resistant bugs and weeds.

(2) Discuss the idea that genetic modification is just an extension of the strategies, such as selective breeding, that farmers have used to modify food crops for centuries.

3 Identify what you think is the strongest argument for GM foods and highlight it in green.

4 Identify what you think is the strongest argument against GM foods and highlight it in red.

Social and ethical concerns

- Large companies that own the patent (exclusive rights) for the GM plants may be able to monopolise (dominate) the world's food market by controlling the distribution of the genetically modified seeds.
- Using genes from animals in food plants may create ethical or religious concerns. For example, eating traces of genetic materials from pork in a vegetable or fruit could be a problem for some religious groups or vegetarians.
- Some people believe that genetically modifying plants and animals is 'playing God' or is unnatural. They say that genes from unrelated species should not be mixed.

(5) A gene from arctic fish has been inserted into tomato plants to help them survive frost damage. Discuss whether a vegetarian should feel concerned about eating such a tomato containing fish genes.

Labelling genetically modified food

In Australia, GM foods and ingredients must be identified on labels with the words 'genetically modified'. GM foods with altered characteristics such as increased nutrient levels, or that need to be cooked or prepared in a different way, also have to be labelled.

Below are two examples of labels for food products:

1 Ingredients: meat (60%), reconstituted textured soy protein*, water, wheat flour, soy protein*, dehydrated potato, salt, beetroot powder, onion powder, mineral salts (450), black pepper, soy lecithin*.
*Genetically modified

2 Ingredients: wheat flour, water added, yeast, soy flour (genetically modified), vegetable oil, sugar, emulsifiers (471, 472E), preservative (282), enzyme amylase.

If the food is unpackaged (for example, loose vegetables), then the information must be displayed with them.

2.10 Genetically modified food

Antibiotic or herbicide resistance

When food crops are genetically modified, scientists introduce a marker gene along with the desired gene. Marker genes are often genes for antibiotic or herbicide resistance.

If the antibiotic resistance genes in the GM food were taken up by bacteria in the human gut, this could reduce the effectiveness of antibiotics given to patients to treat infections. For this to happen the marker gene would have to remain intact after digestion and a long chain of events would have to occur before the antibiotic resistance gene became part of the genetic material of the gut bacteria (see Figure 2.10.1). Each step along the pathway may or may not occur, therefore it is very unlikely that antibiotic resistance becomes part of the bacterial genome.

Many bacteria have naturally occurring antibiotic resistance and these bacteria are in the foods we eat.

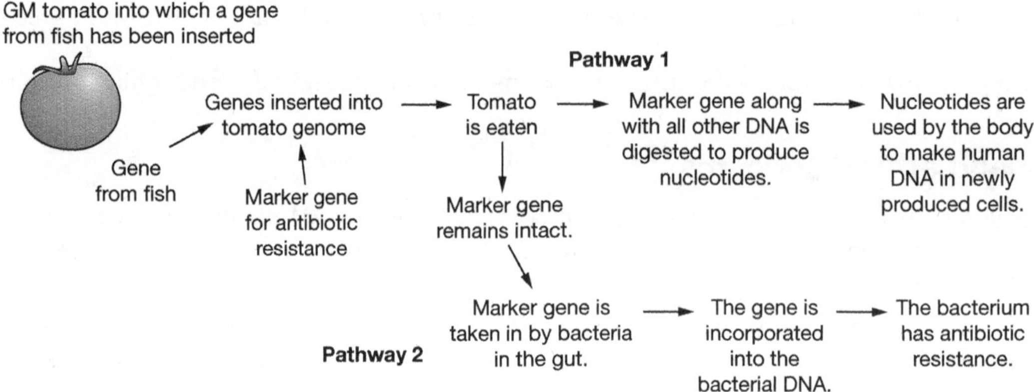

Figure 2.10.1 Pathway 1 is the most likely series of events after a genetically modified tomato (or any other genetically modified food) has been consumed. All the steps in Pathway 2 must occur for antibiotic resistance to become part of the bacterial genome.

6 Decide whether you support the continued use of GM foods in Australia. Justify your opinion.

RATE MY UNDERSTANDING
Shade the face that shows your rating

2.11 Bioinformatics

Science understanding

FOUNDATION | STANDARD | **ADVANCED**

Bioinformatics is a field of science in which biology, information technology and computer science all work together. Analysis of the human genome would be impossible without very powerful computers.

Analysis of the human genome produced long lines of the letters A, T, C and G such as those shown in Figure 2.11.1.

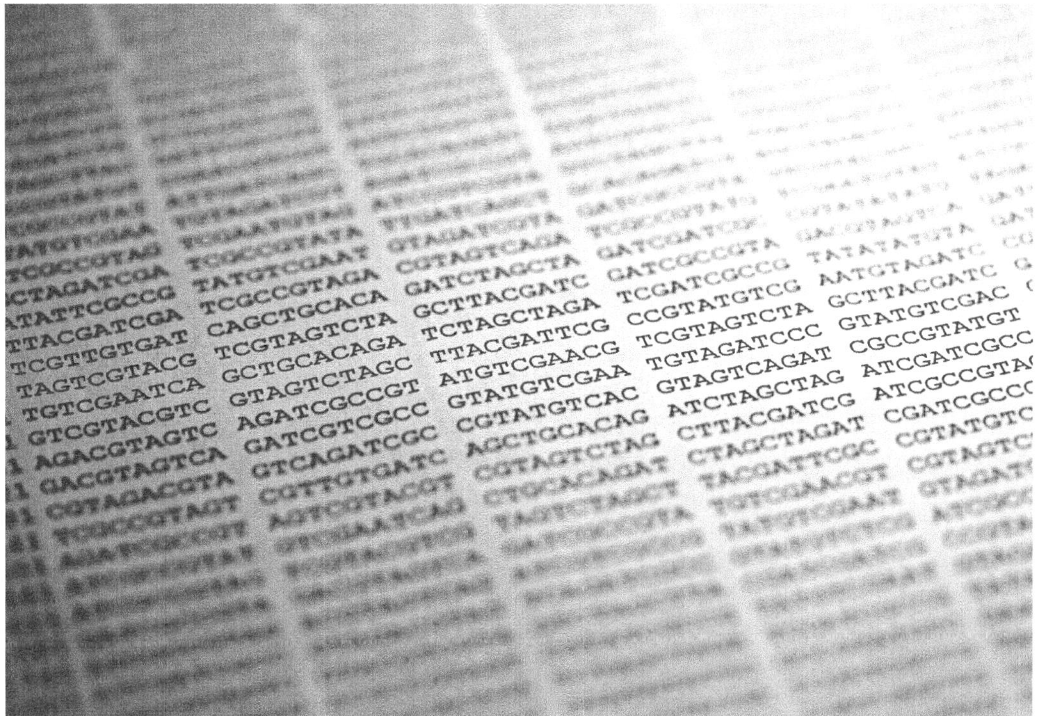

Figure 2.11.1 One way of presenting the data collected in the human genome project is as a list in order of the bases found in the genes.

Scientists discovered that there is one base different between a person with normal haemoglobin in their red blood cells and a person with sickle cell anaemia—a disease that can be fatal.

As part of the gene, a person with normal haemoglobin has the base sequence:

CTG ACT CCT GAG GAG AAG TCT

A person with sickle cell anaemia has the base sequence:

CTG ACT CCT GTG GAG AAG TCT

1 Below is a section of the haemoglobin gene:

TAT ATT CCA AAT AGT AAT GTA GTA CTA GGC AGA CTG TGT AAA
GTT TTT TTT TAG TTA CTT AAT CTG ACT CCT GTG GAG AAG TCT
GTA TCT CAG AGA TAT TTC AAT GTA GTA CTA GGC AGA CTG TGT
AAA GTT TTT TTT TAG TTA CTT AAT CTG ATT CCT

Decide whether the person with this gene suffers from sickle cell anaemia or has normal haemoglobin. Justify your response.

____

____

2.11 Bioinformatics

2 One way of deciding how similar organisms are, is to compare the frequency with which the different bases appear in the same gene such as the gene for haemoglobin.

(a) Identify the number of times each of the four bases appears in the section of the haemoglobin gene shown in question 1.

____

(b) Compare your answer with the totals reached by others in your class.

____

3 Consider the amount of time it took you to work out the answers to questions 1 and 2.

Consider the benefits of using computers for these tasks in terms of:

(a) accuracy

____

(b) time taken.

____

4 Assess whether analysis of the human genome would be possible without computing power given that you analysed bases equivalent of less than 0.00012% of the haemoglobin gene and the haemoglobin gene is one five-thousandth of the human genome.

____

____

____

____

RATE MY UNDERSTANDING
Shade the face that shows your rating

2.12 Personalised medicine

Science understanding

FOUNDATION | STANDARD | **ADVANCED**

Traditionally, doctors have used an individual patient's symptoms along with their medical and family history to diagnose and treat disease. With advances in genetics, the medical professionals now have a more detailed understanding of the role of genetics in disease. There is greater opportunity to tailor treatments to individual needs. In 2012 the Ian Potter Centre for Genomics and Personalised Medicine was established in Melbourne. It is the first personalised medicine division in Australia.

Not all patients with the same disease respond to particular medication in the same way as shown in Figure 2.12.1. Some patients respond well and get better quickly. Others have little or no response, and patients in a third group have a negative response and are harmed by the medicine that was supposed to cure them. Research is ongoing into the genetic reasons for these different responses. It is hoped that in the future, genetic testing of individuals will identify which patients could have a negative reaction to certain medicines. A different treatment would then be given to those patients.

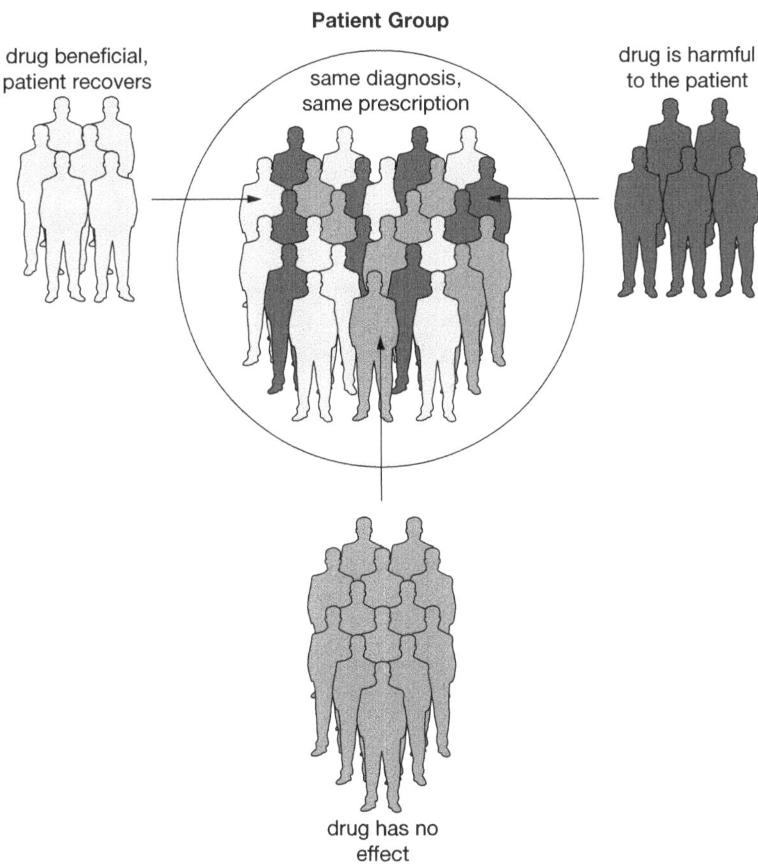

Figure 2.12.1 How you respond to some medicinal drugs depends on your genotype.

Cancers are diseases that have a strong genetic component. The genome of the cancerous tumour is different from the genome of the patient with the disease. By identifying the differences, doctors can make better decisions on the best treatment for a particular cancer. Technology today makes it possible to determine the genome of both the patient and the tumour, but this process takes time and money. However, a small change in the amount of information available allows patients to be put into groups according to their likely response to a particular treatment. This makes the treatment quicker and more likely to be effective. Cancer treatments are expensive, so money is saved getting the treatment right the first time.

2.12 Personalised medicine

Researchers are not sure how much environmental factors such as diet and exercise influence the development of many types of cancers. One way of studying this is to use genetic information to identify people who have alleles predisposing them to a disease and screening these people regularly for the disease. Where possible, prevention programs can be developed that may prevent or delay the onset of disease in individuals at highest risk. If the disease does develop then it can be diagnosed early, when hopefully it will be more treatable.

Some people do not want to know the details of their genetic information such as whether they may have the allele of a gene that is linked to heart disease or to breast or prostate cancer. However, research shows that having the allele may not change the chance of developing the disease by very much. For example, there is a gene linked with breast cancer that increases the chance that tumours will develop. Some individuals with that gene also have a gene that acts to stop tumours forming. In this case the disease is less likely to develop. Research has shown that the majority of breast cancer cases are not linked to identified genes.

1 Explain how knowing a person's genome could reduce the chances of that person having an adverse reaction to a prescribed medicine.

____

____

____

2 **(a)** Describe the information doctors collect before deciding how to treat a cancer patient.

____

____

(b) Explain how this benefits the patient.

____

____

____

3 Explain how knowing that you are 'at risk of developing a particular disease' could help you stay healthy.

____

____

____

RATE MY UNDERSTANDING
Shade the face that shows your rating

2.13 Assisted reproduction

Science understanding

FOUNDATION	STANDARD	ADVANCED

Use the information provided on the noteboard in Figure 2.13.1 and your knowledge of genetics to answer the questions that follow.

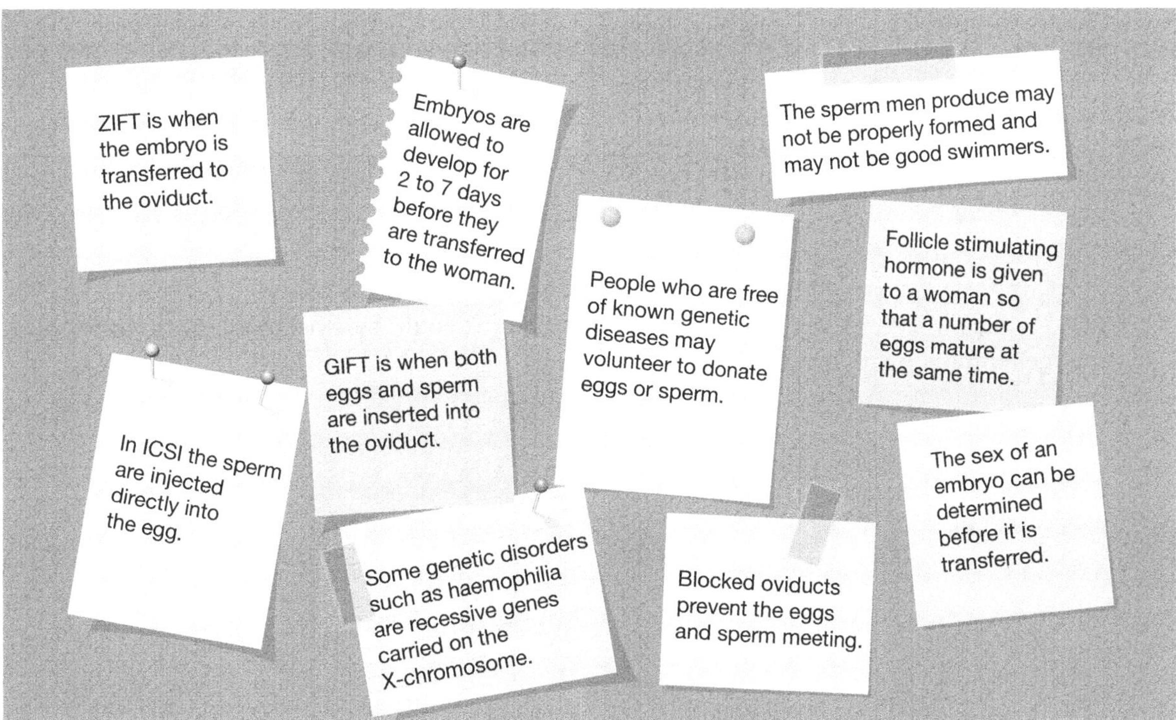

Figure 2.13.1

1. Name the hormone that causes eggs to mature. ______________________

2. State two causes of infertility.

____

____

3. Identify the assisted reproduction technique that would be used if:

   (a) the sperm were not strong enough to enter the egg to fertilise it

   ____

   ____

   (b) the sperm were not good swimmers.

   ____

   ____

2.13 Assisted reproduction

4 (a) Explain why zygote intrafallopian transfer (ZIFT) would not work when the woman has blocked oviducts.

(b) Explain how assisted reproduction could be used to ensure that the child of a carrier of haemophilia was not a haemophiliac.

(c) Deduce whether a child born using this technique could pass haemophilia on to the next generation. Justify your answer.

RATE MY UNDERSTANDING
Shade the face that shows your rating

2.14 Literacy review

Science understanding

FOUNDATION	STANDARD	ADVANCED

1 Use your knowledge of genetics and DNA by using the words from the box to complete the sentences below. Words may be used more than once.

adult stem cells	alleles	complementary base pairs	cytosine	deoxyribonucleic acid	
differentiate	gene splicing	genetically modified	genotype	homologous	meiosis
mitosis	phenotype	pluripotent	recombinant DNA	replication	thymine

(a) ________________ ________________ (DNA) is the complex molecule that carries the genetic code.

(b) The four nitrogen-rich bases pair up as c ________________ b ________________ p ________________. Adenine pairs with ________________ and guanine pairs with ________________.

(c) DNA is able to make copies of itself in a process known as ________________.

(d) DNA replication takes place before both types of cell division, which are called: ________________ and ________________.

(e) ________________ produces two daughter cells that are identical to the parent cell.

(f) ________________ produces gametes (eggs and sperm) that have half the number of chromosomes of the original cell.

(g) ________________ chromosomes have the same genes for particular characteristics at the same location on the chromosome.

(h) Variations of genes are known as ________________.

(i) The ________________ is the actual genetic information carried by an individual. The ________________ is the observable characteristics of the individual.

(j) During the process of growth and maturation, cells ________________, meaning they become different from each other in structure and function.

(k) When the genetic information contained within the nucleus is changed by inserting new genes, the cell has been ________________ ________________.

(l) Scientists use ________________ ________________ to remove unwanted genes and add new genes into the DNA of the bacterium. The product is ________________ ________________, which is DNA that has been recombined with other genes.

(m) ________________ ________________ ________________ are the cells that allow you to regenerate and repair your tissues.

(n) Embryonic stem cells are ________________. They are capable of becoming any one of the 220 different cell types found in the human body.

RATE MY UNDERSTANDING
Shade the face that shows your rating

2.15 Thinking about my learning

1 Imagine that you are going to be presenting a talk to a Year 8 class about genetics and inheritance. There will be time for questions at the end. Some questions that you could be asked are given in the table below. Consider how confidently you could answer them. Tick the appropriate box in the table.

Question	I could answer this question with confidence	I think I could answer this question	I don't think I could answer this question
1 What is a double helix?			
2 What did Watson and Crick do that was special?			
3 What is DNA made of?			
4 What are genes?			
5 Where do you find genes?			
6 What is the process of making new body cells called?			
7 What sorts of cells are made during meiosis?			
8 How many chromosomes does a human have?			
9 What is meant by a dominant trait?			
10 My mum has black hair and my dad has blonde hair—so why is my hair red?			
11 What is a Punnett square?			
12 What is an allele?			
13 What is a mutation?			
14 Are all mutations the same?			
15 Are mutations all bad?			
16 What is genetic engineering?			
17 What is a plasmid?			
18 How are plasmids used for genetic engineering?			
19 What is the Human Genome Project?			
20 What advantages have been gained from the Human Genome Project?			
21 What is a disadvantage of genetic testing?			
22 How is a DNA profile made?			
23 Why can a DNA profile distinguish between different people?			

2 Write down the numbers of the questions which you can answer now but which you didn't know the answer to before you started this unit.

____

3 What do you think was the most important thing that you learned in this topic?

____

4 What was the most interesting thing that you learned in this topic?

____

CHAPTER 3

Geological time

3.1 Knowledge preview

Science understanding

FOUNDATION | STANDARD | ADVANCED

1. Life on Earth has changed over time. Some important events in the Earth's history are listed below. Number the events from most ancient (1) to most recent (14).

______ mammals dominate	______ first land plants	______ first photosynthetic organisms
______ first reptiles	______ dinosaur extinction	______ first primates
______ fish appear	______ jellyfish first seen	______ the Chicxulub meteor arrives
______ first cells	______ first mass extinction occurs	______ first land vertebrates
______ humans appear	______ mammals evolve	

2. Much of what we know about life in the past comes from fossils. Identify which of the following are fossils. Write either yes or no beside each item.

frozen mammoth	________	dinosaur footprint	________
insect in amber	________	burrow of an extinct mole	________
the pyramids	________	cast of a 50-million-year-old jellyfish	________
pre-historic axe	________	shell from a 200-million-year-old mussel	________

3. The decay of radioactive materials is used to date some fossils. Carbon-14 (^{14}C) is one radioactive material used to date some fossils.

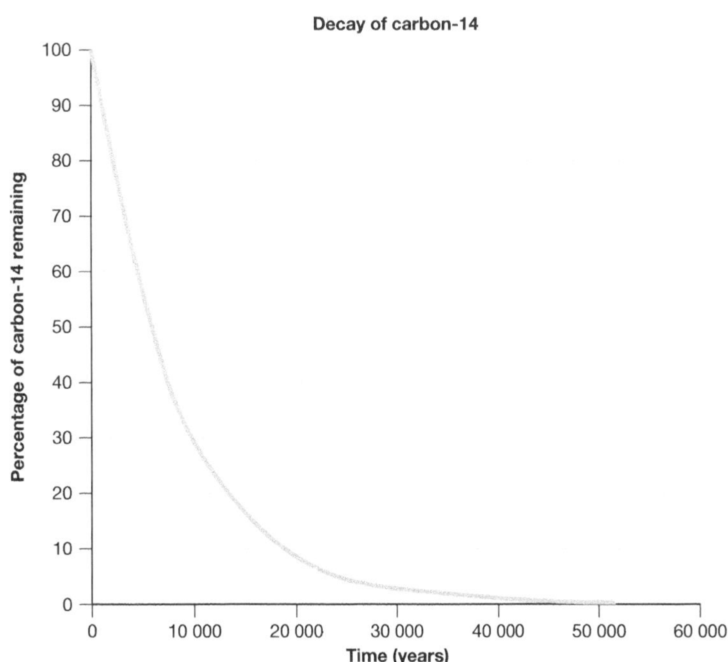

3.1 Knowledge preview

Use the graph on page 33 to:

(a) determine the age of a fossil having 50% of carbon-14 (^{14}C) remaining ________

(b) determine how much ^{14}C would be left in a fossil that is 23 000 years old. ________

4 Throughout Earth's history there have been many times when species have become extinct.

(a) What is extinction?

____

____

(b) How many different causes of large-scale extinctions do you know about? Describe them.

____

____

____

____

____

____

RATE MY UNDERSTANDING
Shade the face that shows your rating   

3.2 Identifying fossils

Science inquiry skills

FOUNDATION | **STANDARD** | ADVANCED

Each of these photos shows a different fossilised organism.

A

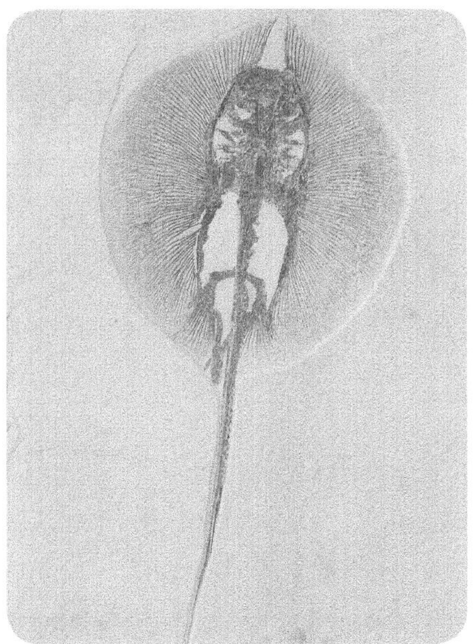

B

C

D

E

F

3.2 Identifying fossils

G

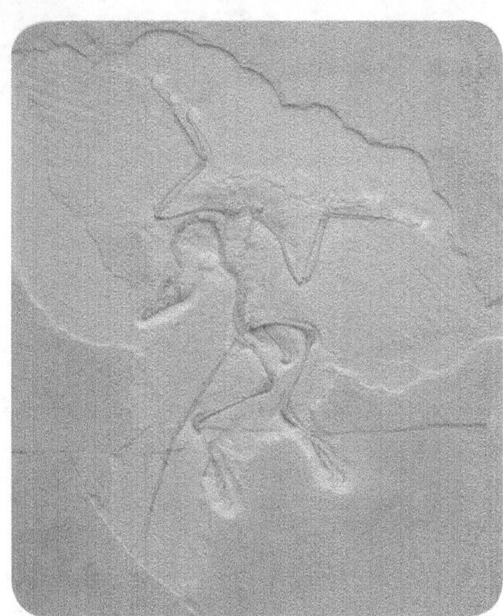

H

I

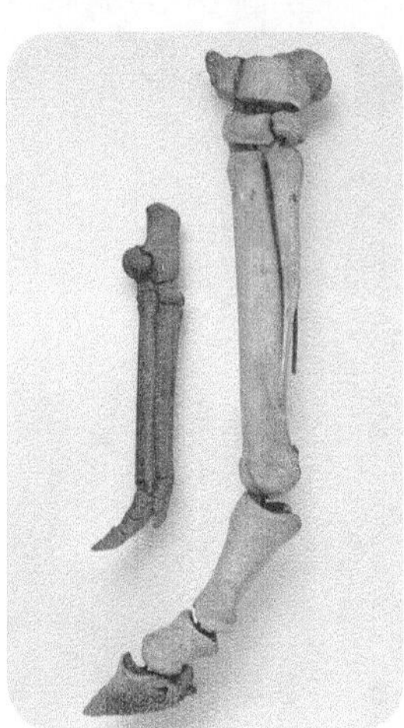

J

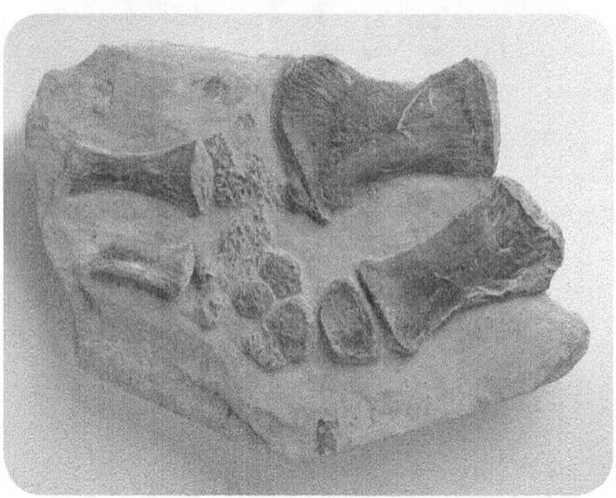

1. Identify each type of organism that has been fossilised in the photos by using a general name such as 'cat' or 'snake'.

   A ______________________ B ______________________

   C ______________________ D ______________________

   E ______________________ F ______________________

   G ______________________ H ______________________

   I ______________________ J ______________________

2. Identify by letter a fossil that is probably a carbon film. ________
3. Identify the six vertebrate fossils by writing their letters.

   ____

4. Name some types of fossils that are not shown in the photos.

   ____

   ____

silica (*n*) a mineral
tar pit (*n*) a hole filled with thick, sticky oil

5 Fossil J is a replacement fossil in which the bones have turned to silica, whereas fossil D is an original fossil in which the animal has been preserved by falling into a tar pit.

(a) Explain some differences in the way fossils J and D were formed.

____

____

____

(b) What information could you obtain from fossil D about its original body chemicals that you could not obtain from fossil J?

____

____

____

6 Fossil G is one of the most famous fossils ever found. It belongs to an animal called *Archaeopteryx*. When it was first discovered, palaeontologists thought it was a dinosaur. But along with the bones, something else was fossilised. Name and describe this other feature you can see in the fossil.

____

____

____

____

7 Explain what is meant by the *fossil record*.

____

____

____

RATE MY UNDERSTANDING
Shade the face that shows your rating

3.3 Relative dating

Science inquiry skills

FOUNDATION | **STANDARD** | ADVANCED

Processing & Analysing | Communicating

Some geologists were investigating rock strata in a gently sloping river valley. They drilled down into the rocks in four places labelled A, B, C and D, shown in Figure 3.3.1. This diagram also shows the ground contours in 5-metre intervals. A contour line shows all places that are the same height above sea level. For example, the 300 line shows all places that are 300 metres above sea level.

index fossil (*n*) a fossil that can be used to compare the ages of rock strata

strata (*n*) (plural; singular is stratum) layers, levels

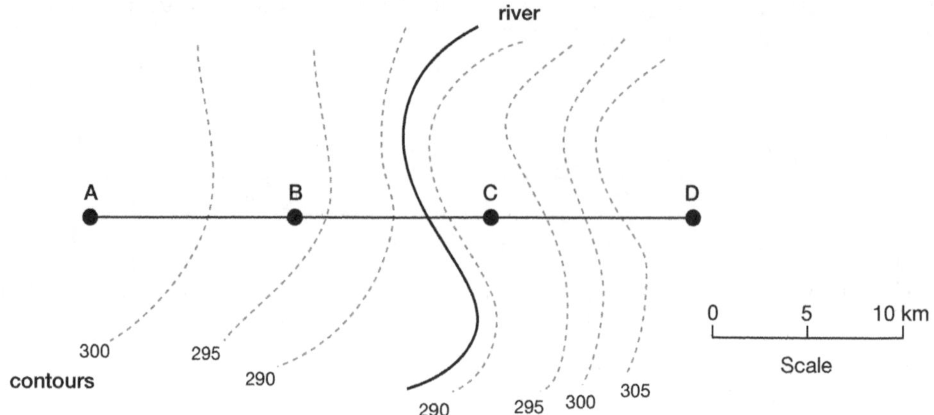

Figure 3.3.1 Position of drilling sites A to D shown on a contour map

The rocks from the four sites, such as limestone or shale, were carefully studied for index fossils. The palaeontologists found that five index fossils occurred across the four sites. You can see these five index fossils in Figure 3.3.2.

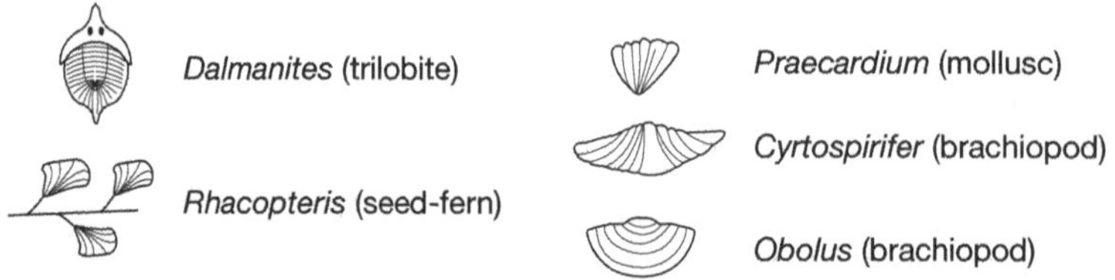

Figure 3.3.2 Index fossils found at drilling sites A to D

1 Go to page 47 and cut out the top diagram showing the different strata found in the four sites

Cut out each of the 4 strata diagrams and arrange A to D from left to right. Move each column up or down until you have lined up the same layers from each site.

To do this task, you need to know that:

- the rock layers at any site are numbered from the top layer down
- the same numbers do not mean they are the same layers; for example, A1 is not the same as B1
- a layer in a particular site may not have an index fossil present even though it may be the same age as the layer in another site. There is never a guarantee that a layer at a particular site will have fossils in it.

When you are sure that your strata are correct, glue your arrangement of the four columns onto the space provided at the top of the next page.

(2) Explain how the contour lines in Figure 3.3.1 provide extra information to support your arrangement in question 1.

(3) Name the process that you carried out in question 1.

(4) **(a)** Identify the oldest layer (by site letter and layer number) over the four sites.

(b) Why is this layer the oldest?

(5) **(a)** Identify the youngest layer (by site letter and layer number) over the four sites.

(b) Why is this layer the youngest?

(6) Identify a layer (by site letter and layer number) that may have been expected to have *Dalmanites* fossils in it, but which did not have any fossils.

(7) Which two layers are the same age as A2?

RATE MY UNDERSTANDING
Shade the face that shows your rating   

3.4 Absolute dating

Science inquiry skills

FOUNDATION | **STANDARD** | ADVANCED

Processing & Analysing

In this activity, you will use a graph called a decay curve to find the age of some fossils. A decay curve shows how the level of radioactivity in a rock decreases over time.

Potassium–argon dating

Radioactive potassium-40 breaks down into a gas called argon, which becomes trapped in the rock. The amount of potassium decaying into argon over time can be measured. A graph called a decay curve is then drawn, which shows how the proportion of potassium-40 changes over time. The decay curve for potassium-40 is shown below.

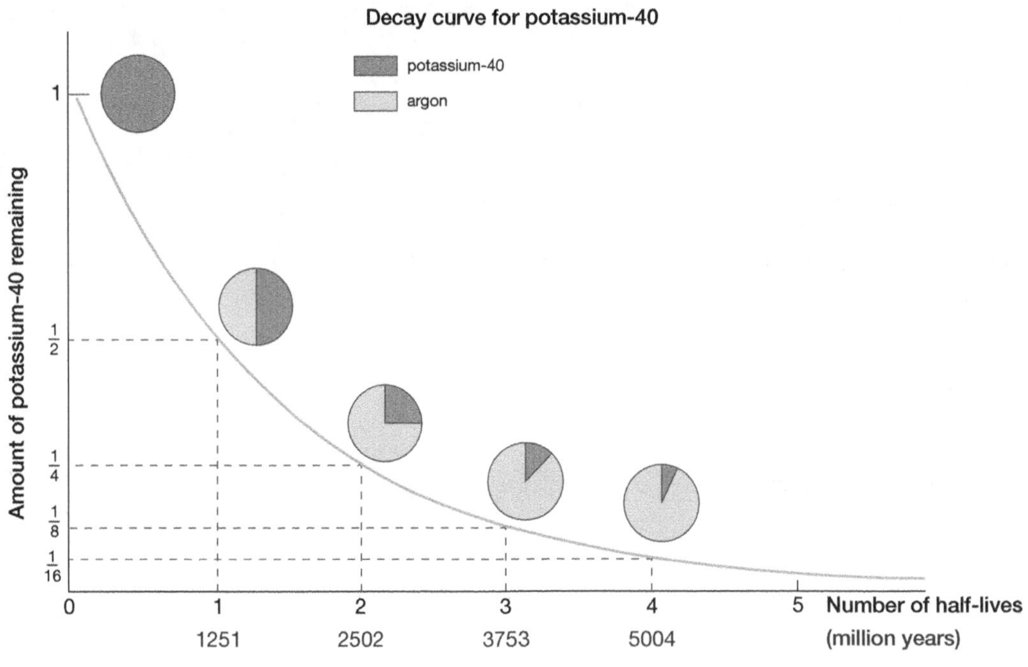

To determine the age of a rock, scientists measure the amount of argon and potassium-40 in a sample. They then calculate the proportion of the original potassium-40 that is left. Use the graph above to answer the following questions.

dating (*n*) finding the age of something
half-life (*n*) the time it takes for something to reduce by half

1. How old is a rock that has half of its original potassium-40 left?

___

2. How old is a rock that has three-quarters of its potassium-40 left?

___

3. How many half-lives and years would pass before the rock had only one-quarter of its potassium-40 left?

   Half-lives: ___ Years: ___

4. The Earth is estimated to be 4500 million years old. How many half-lives of potassium-40 have there been in this time?

___

RATE MY UNDERSTANDING
Shade the face that shows your rating   

3.5 The move to land

Science understanding

FOUNDATION | **STANDARD** | ADVANCED

Scientists have concluded from the fossil record that life began in the sea, and then moved to the land during the Devonian period. The vertebrate fossil record shows at least 10 species of Devonian lobe-finned fish and early amphibians that provide evidence to support this hypothesis.

The main source of evidence for this movement from sea to land is in the structure of the limbs. Tetrapods (four-footed land animals such as amphibians, reptiles and mammals) walk on the land supported by the bones in their legs. So if some fish developed into land animals, you would expect to find fish limbs that had similar bones in them to the tetrapods. This is exactly what many of the fossils, such as those in Figure 3.5.1, show, especially in the front limbs.

amphibian (*n*) an animal that can live in water and on land

common (*adj*) a shared characteristic

digit (*n*) a finger or toe

interpret (*v*) to give meaning to information

limb (*n*) part of the body used for movement such as leg, arm or fin

The humerus, radius and ulna are common limb bones in amphibians, reptiles, birds and mammals. You have these bones in your arm. Lobe-finned fish had bones at the base of their fins.

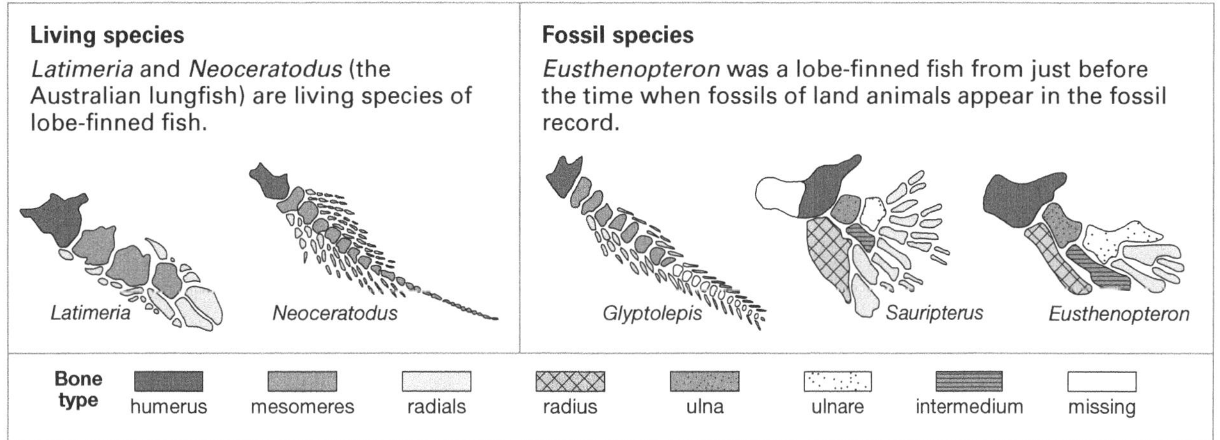

Figure 3.5.1 The front limbs of living species (left) and fossil species (right) of lobe-finned fish. Common bones are shaded the same.

To understand how tetrapod and lobe-finned fish limbs compare, consider the fossils in Figure 3.5.2 on page 42. *Tiktaalik* and *Panderichthys* were more like fish than amphibians. *Tiktaalik* had a wrist-like structure in its limbs. Scientists interpreted this feature as being able to support the weight of the animal for brief periods. The other four species were amphibians that are classified as tetrapods, and clearly had fingers and toes. *Acanthostega* had eight digits and strong limbs more like a land animal's leg than a fish fin. However, *Acanthostega*'s bones could not have supported the weight of the animal's body on land for long periods of time. *Tulerpeton*'s limbs were similar to those of *Acanthostega*. *Proterogyrinus* and *Limnoscelis* had limbs that would support them on land for longer periods than earlier amphibians. They may have lived both in water and on land.

3.5 The move to land

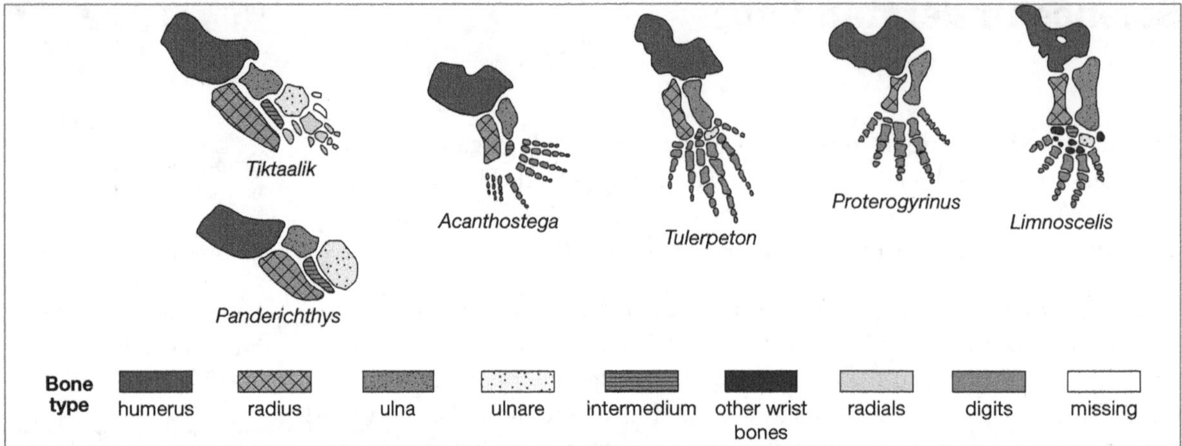

Figure 3.5.2 Some early lobe-finned fish and amphibians and their limb structure

In considering how the change may have occurred in fish fins, scientists considered the environments in which early lobe-finned fish probably lived. Scientists wondered why lobe-finned fish and early amphibians developed bony fins and limbs, when they were living in water rather than walking on land. One possibility is that these limbs developed in species living in freshwater swamps where there was much vegetation. Bony supports in limbs would have been much better than fins for moving through the vegetation by gripping it and pushing on it. The lobe-finned fish and early amphibians may also have been able to move brief distances over land from pool to pool, dragging themselves using their front limbs.

To be sure about how this change from water to land occurred, more fossils are needed, especially fossils that clearly show the early beginnings of digits (fingers and toes).

1 Name the geological period in which the fossil evidence suggests vertebrates moved from the sea to the land.

2 Define what is meant by a *tetrapod.*

(3) Why are palaeontologists especially interested in the limb structure of lobe-finned fish and early amphibians?

(4) Name the limb bone that is common to lobe-finned fish and early amphibians.

5 Name three bones that were found in all early amphibian fossils.

(6) What advantages did bony limbs probably give these lobe-finned fish and amphibians?

RATE MY UNDERSTANDING
Shade the face that shows your rating

3.6 Dinobirds

Science understanding

FOUNDATION | **STANDARD** | ADVANCED

Biologists have concluded that birds are related to dinosaurs. The idea was first proposed in the 19th century just after the bird-like fossil skeleton *Archaeopteryx* was discovered. The skeleton showed *Archaeopteryx* to be almost identical to a dinosaur called *Compsognathus*. *Archaeopteryx* had some features not found in birds, such as teeth, a long tail with bones in it and claws on its front legs. However, like a bird, *Archaeopteryx* had a 'wishbone' in the chest (necessary for flight) and feathers. The feathers were present as external moulds in the rock surrounding the fossilised bones.

adaptation (*n*) characteristic that gives a greater chance of surviving and reproducing

ancestor (*n*) previous generations, parents, grandparents etc

bipedal (*adj*) walking on two legs

descendants (*n*) later generations, children, grandchildren etc

genera (*n*) plural of genus; group including more than one species

sac (*n*) small pocket

Fossils in China

In the 1990s, at Liaoning in China, many fossils of different dinosaurs were discovered that had strange-looking furry material covering parts of the body. The fossils, such as the one shown in Figure 3.6.1, came from an area that had regularly been covered by volcanic ash, which preserved the fossils in beautiful detail. This method of preservation also enabled an accurate dating to be made of the age of the fossils. They were 124 million years old.

The furry covering turned out to be preserved body tissue and some of it was original fossils of the animal's body proteins. As more fossils of different species were discovered, it became obvious that these structures were simple feathers. The feathers were of different forms, and some were like modern bird feathers. Chemical tests on the fossilised feathers of one dinosaur showed their chemical structure to be like that of modern bird feathers.

Figure 3.6.1 A feathered dinosaur from Liaoning, China. The feathers are the dark lines at the back of the dinosaur's limbs.

Dinosaurs with feathers

The dinosaurs that had feathers were mainly from one group called theropods. Within the theropods, the main feathered dinosaurs belonged to the Coelurosaurs. Most Coelurosaurs were bipedal predators (they walked on two legs). They had long tails and their 'arms' had an ulna bone that was curved outwards, just like modern birds. Some fossils show what look like air sacs in their breathing systems, also similar to many modern birds. Coelurosaurs also laid eggs and nested in a similar way to birds. Fossils have been found of adult Coelurosaurs sitting on a nest of eggs.

Today we know that there are more than 20 genera of dinosaurs that had feathers, nearly all of which are theropods. There are so many similarities in the skeletons of theropod dinosaurs and birds that most palaeontologists are convinced that theropods were the ancestors of birds.

3.6 Dinobirds

Recent evidence

The latest find to support the hypothesis that birds are descended from dinosaurs is a dinosaur called *Anchiornis* discovered in 2009. It has been dated at 161 to 151 million years old from the Jurassic period.

This makes it older than *Archaeopteryx*. It was completely covered in feathers, some like those used in flight in modern birds. It also had long feathers on its legs. Although the arms of *Anchiornis* were quite long, it probably could not fly very well. On current evidence, biologists have concluded that dinosaur feathers were not an adaptation for flight. They think that feathers were for temperature control, and probably helped to keep some dinosaurs warm. It was only millions of years later that feathers enabled descendants of these dinosaurs to fly.

1 What have palaeontologists concluded about birds and dinosaurs?

____

____

2 Name the first dinosaur discovered that had bird-like features.

____

3 What bird-like features have palaeontologists identified in *Archaeopteryx*?

____

____

(4) What information did the Liaoning fossils provide about the feathers of dinosaurs that the *Archaeopteryx* fossil did not?

____

____

(5) Describe some features shared by Coelurosaurs and birds.

____

____

(6) Explain the importance of *Anchiornis* fossils.

____

____

7 How may feathers have been an advantage to the early dinosaurs?

____

____

____

RATE MY UNDERSTANDING
Shade the face that shows your rating

3.7 Constructing dinosaurs

Science inquiry skills

FOUNDATION | **STANDARD** | ADVANCED

Processing & Analysing | Communicating

The second diagram on page 47 shows mixed-up bones from some different dinosaurs.

1. Try to construct these three dinosaur skeletons. Cut out all the bones and arrange them in the space below to show the skeletons as you think they were in these dinosaurs. Move the bones around until you are sure they are correct, then glue them onto the space below.

2. What are the names of the dinosaurs in question 1?

3. Discuss what methods you used to decide how the bones fitted together.

RATE MY UNDERSTANDING
Shade the face that shows your rating

This page has been left blank

This diagram of rock strata is to be cut out and used for **Worksheet 3.3**.

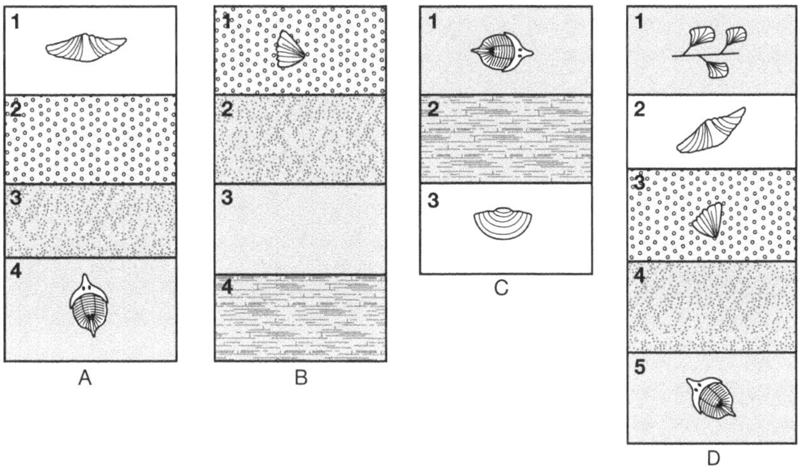

This diagram of dinosaur bones is to be cut out and used for **Worksheet 3.7**.

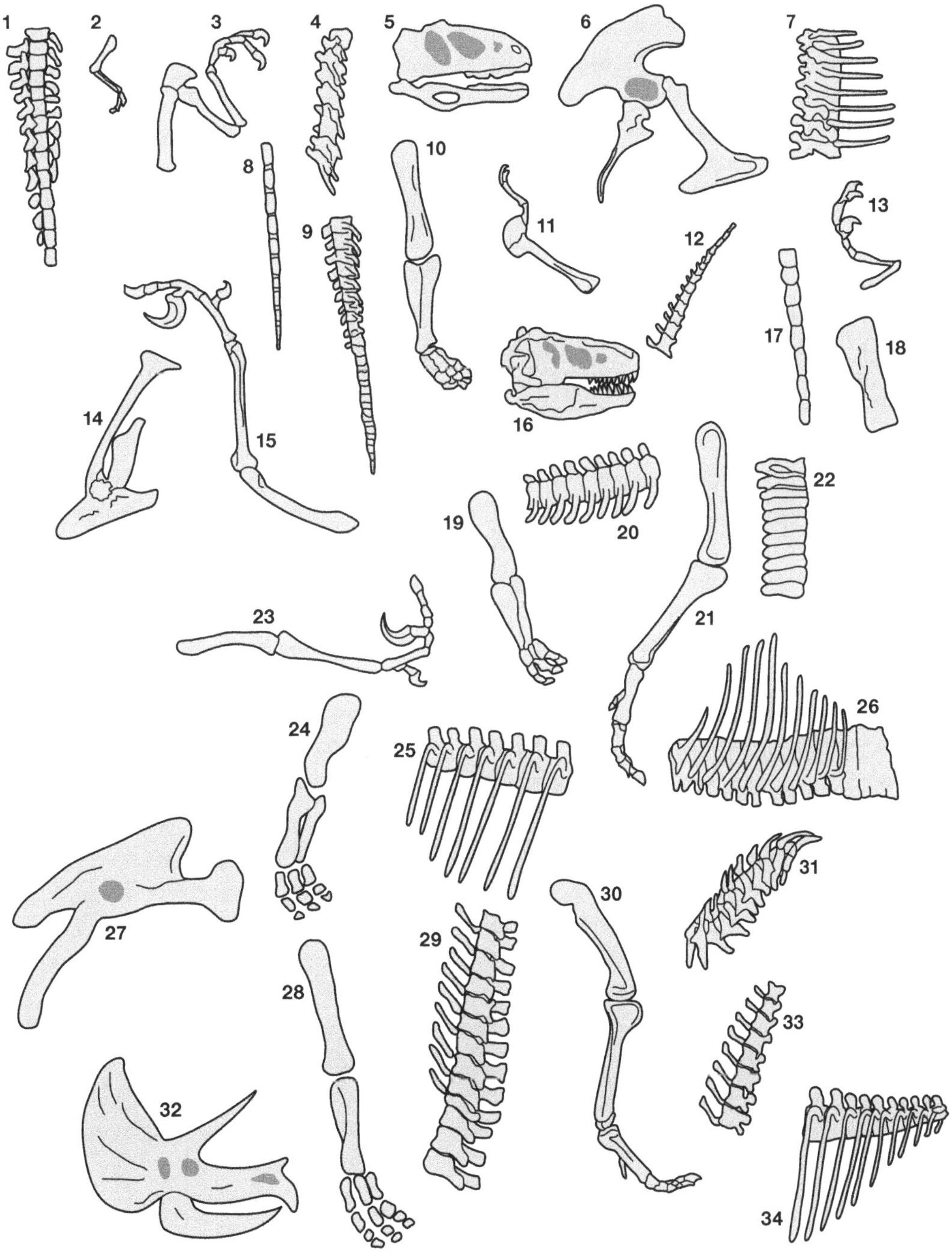

This page has been left blank for the cut-out activities on page 47

3.8 Literacy review

Science understanding

FOUNDATION	STANDARD	ADVANCED

1 The table below lists terms and definitions about geological time. Some of the terms and definitions are missing. Complete the table by filling in the missing information.

Terms	Definitions
fossil	
	scientist who studies prehistoric life
	fossil that can be used to compare the relative age of strata in different locations
stratigraphy	
lobe-finned	
	land animals with four limbs, including amphibians, reptiles, birds and mammals
	land animals in the reptile group called archosaurs but which also had their limbs placed vertically beneath their bodies
fossil record	
relative	

2 Match the term with its correct definition by drawing a line between them.

Term	Definition
carbon film fossil	minerals in part of a fossil, such as its skeleton, are lost and replaced by another mineral
cast	an imprint of the outside of the body in rock
mould	when an organism in rock decomposes and the space in the rock fills with soil that turns to rock
palaeontology	a scale showing the history of life and geology
absolute dating	layers of sedimentary rock or soil
half-life	when a dead body partially decays and leaves a thin black deposit of carbon
strata	a way of determining the actual age of rocks and fossils
replacement fossil	the time it takes for half of a radioactive sample to decay
geological time	the study of prehistoric life

RATE MY UNDERSTANDING
Shade the face that shows your rating

3.9 Thinking about my learning

1 Some of the ideas around geological time need to be memorised, that is, they must be known. Other ideas are more complex and they must be understood as well as known. Consider whether each of the ideas listed in the table below is one you must just know (memorise) or whether they are ideas you must know as well as understand.

Put a 'K' for know or a 'KU' for know and understand in the first column of the table. Then you should put a tick in the column which best describes how you feel about your knowledge/ understanding of each idea.

Idea	K or KU	I am still confused about this idea	I feel fairly comfortable that I know/ understand this idea	I am so confident about my knowledge/ understanding of this idea that I could explain it to someone else
what a fossil is				
a list of the different types of fossils				
the ideal conditions for fossilisation				
how radioactive dating is performed				
calculation of a half-life				
use of a decay curve to find the age of a fossil				
what an index fossil is				
what makes a good index fossil				
how to find the relative ages of different rock strata				
the difference between absolute and relative dating				
when the various types of organisms evolved				
the different divisions of the Earth's history				
the causes and timings of mass extinctions				
when significant events in the Earth's history occurred				

2 Consider the ideas about which you are still confused.

(a) What strategies could you adopt to help you to know and/or understand these ideas better?

____

____

(b) How will these strategies help?

____

____

(c) Put your strategies into action then re-consider the table above and move any ticks if you have addressed that difficulty.

CHAPTER 4

Natural selection and evolution

4.1 Knowledge preview

Science understanding

FOUNDATION | **STANDARD** | ADVANCED

1 **(a)** Identify each of the following statements as either true or false. Write your answer in the space provided.

(b) For each statement that you identified as false, state what is incorrect about the statement.

Statement	True / False	What is incorrect?
Antibiotic resistant bacteria have become a serious problem because the antibiotics have caused mutations in the DNA of the bacteria.		
Sharks and dolphins are similar in shape because the environment has selected that shape.		
The human hand, bird wing and horse hoof show that these organisms have a common ancestor.		
Rising salt is a problem around the Murray River. Some plants are being badly affected. It could be expected that during their lifetimes, some individuals will evolve a tolerance to high salt concentrations in the soil.		
A ship runs aground on a small island and some rats are left behind on the island when the ship is salvaged. The surviving rats have to eat the tough vegetation and their cheek muscles become enlarged. It could be expected that future generations of rats on the island will be born with large cheek muscles.		
A farmer started using a new pesticide on his crops and as a result most insects died. It would be expected that in a few years a population of resistant insects will have developed.		
Earth's climate has been warming ever since the last ice age. In response to the rising temperatures, some species developed mutations to ensure their survival.		
There is a swamp near a town in a tropical area. Mosquitoes which transmit malaria live in that swamp. The swamp is sprayed four times a year with insecticide to kill the mosquitoes. After 20 years, most of the mosquito population will be resistant to the insecticide.		

4.2 Vertebrate limbs

Science inquiry skills

FOUNDATION | STANDARD | ADVANCED

Processing & Analysing

Vertebrate limbs appear to have a basic underlying pattern, which is called the pentadactyl limb. The word pentadactyl originates from ancient Greek: *penta* meaning five and *dactyl* meaning digits (toes, fingers). This pattern is shown in Figure 4.2.1.

fore (*prefix*) front
generalised (*adj*) approximate, not exact
hind (*prefix*) back
to scale (*adv*) at the correct size in relation to the real object

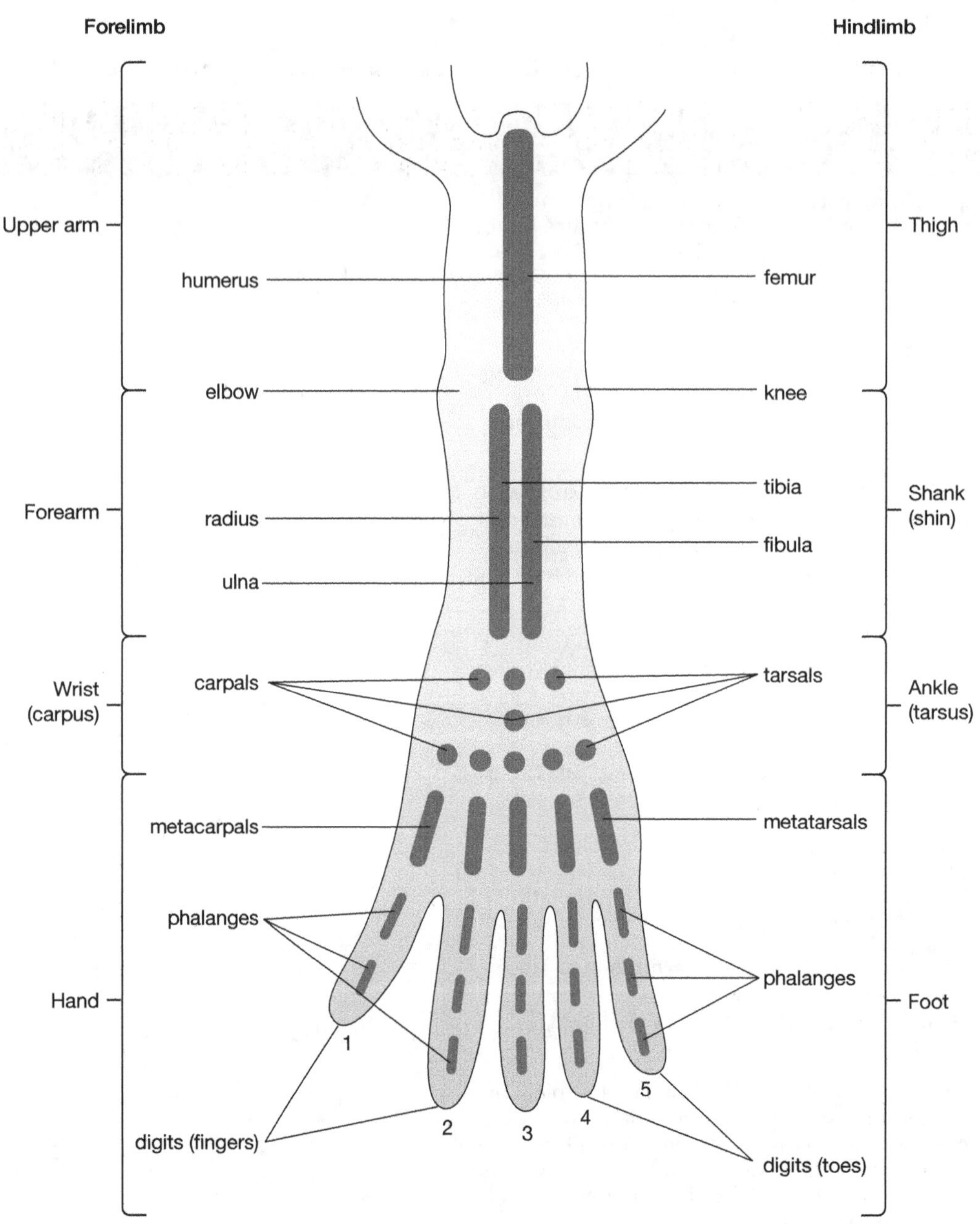

Figure 4.2.1 Generalised pentadactyl limb

4.2 Vertebrate limbs

Some different vertebrate limbs are shown in Figure 4.2.2. These are not drawn to scale.

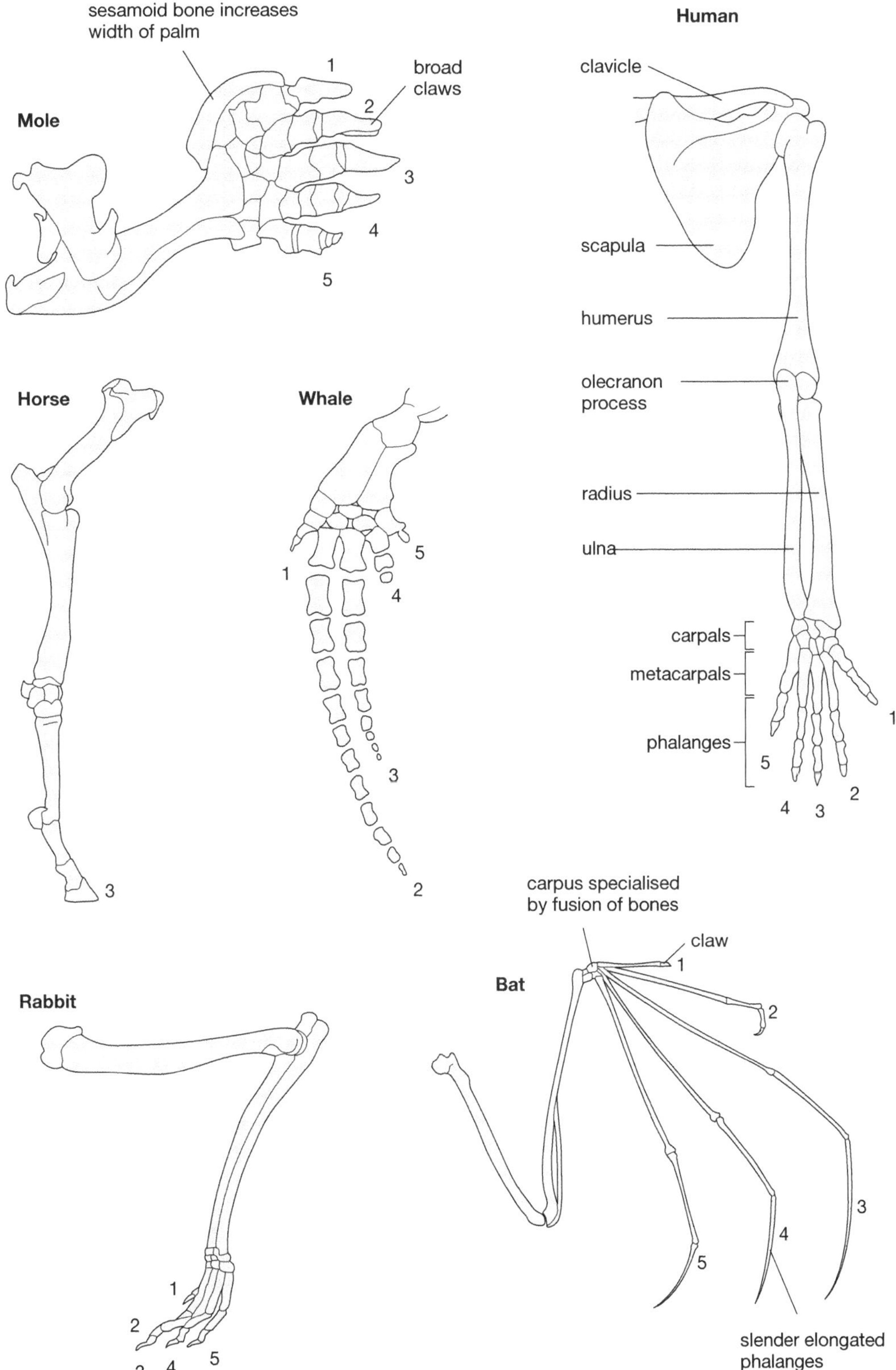

Figure 4.2.2 Variations of the vertebrate forelimb (front limb) structure

4.2 Vertebrate limbs

1. **(a)** Identify four limbs from Figure 4.2.2 that are most like the basic pattern of vertebrate limb that is shown in Figure 4.2.1.

   (b) Explain your choices.

2. **(a)** Identify the two limbs that are least like the basic pattern.

   (b) Explain your choices.

3. Explain how bat, horse and mole limbs are suited to their particular function.

4. **(a)** Identify which animal's limb bone seems to have changed the most from the generalised pentadactyl limb.

   (b) Justify your choice.

5. According to scientific studies, bats are the closest of these animals to humans in terms of ancestry, and rabbits are the second closest. Moles, horses and whales are more distant relatives. Provide evidence to support or refute that, in terms of limb structure, bats and rabbits are more closely related to humans than moles, horses and whales.

RATE MY UNDERSTANDING
Shade the face that shows your rating

4.3 Natural selection of snails

Science inquiry skills

FOUNDATION | STANDARD | **ADVANCED**

In the 1950s, two English scientists, Arthur Sheppard and Philip Cain, investigated natural selection of the brown-lipped snail *Cepaea nemoralis* by the song thrush (a predator of the snail). The scientists studied two areas.

The two areas were a forest floor and an open grass-covered field. The forest floor was covered in brown, dead leaves and often partly in shadow. The open grassy areas were well lit and a yellowish-green colour.

The song thrush preyed on the snails in both areas. The thrush smashed the snail shells open on special rocks, called anvils, which were in the birds' territories. This is shown in Figure 4.3.1. The snails came in two colours—brown and yellow. Some also had black and dark-brown bands. Two different types are shown in Figure 4.3.2.

banded (*adj*) having striped markings

beechwood (*n*) a type of tree

prey on (*v*) to hunt and eat an animal

stable (*adj*) staying the same, not changing

uniform (*adj*) the same all over

Figure 4.3.1 A song thrush eating a brown-lipped snail

Figure 4.3.2 Variations in colour and banding in the brown-lipped snail

Study area 1

At Marley Bog, in a grassy field that has a uniform greenish background, 264 out of 560 living snails collected by Sheppard and Cain in a sample were banded. When the anvils in the area were studied, 486 out of 863 dead snails were banded.

1. For Marley Bog, calculate and compare the percentage of banded snails in the living population with the percentage of banded snails killed by the birds.

____

____

2. Explain how the selection observed in question 1 could change the proportion of the banded genes in the population in this area over a long time.

____

____

Study area 2

The second study area in a beechwood forest floor looked at the effect of seasons on the selection of the snails by the song thrush. In early spring, the beechwood floor is covered in dead, brown leaves. As spring continues, plants sprout and grow, hiding the dead leaves in new green growth. The results obtained in one study are shown in Table 4.3.1. The last column gives the percentage of the population in early and late spring that had a brown shell. Note that, throughout spring, the percentage of living snails found that had a brown shell was fairly stable.

Table 4.3.1 Snail population in a beechwood forest

Season	Number of snails killed—yellow shell	Number of snails killed—brown shell	Total snails killed in season	Percentage of snails killed—brown shell	Percentage of total population alive—brown shell
early spring (brown leaf litter)	15	21	36		76
late spring (green growth)	9	64	73		72

(3) Use the space below to calculate the percentage of snails killed that were brown in early spring and in late spring. Enter your results on the table.

(4) **(a)** Compare the percentage of snails killed that were brown in early and late spring.

(b) Suggest a reason why the percentage of brown snails killed varied from early to late spring.

5 One possibility to explain the data in Table 4.3.1 is that the change in the percentage of brown snails killed was due to a chance increase in the proportion of brown snails. Use the data in the last column of the table to evaluate this explanation.

RATE MY UNDERSTANDING
Shade the face that shows your rating

4.4 Combating bacterial resistance

Science understanding

FOUNDATION | **STANDARD** | ADVANCED

factor (*n*) reason
inactivate (*v*) to stop something being active
mutation (*n*) changes in the genes
widespread (*adj*) found in many places
withstand (*v*) to oppose, to resist

The development of antibiotic resistance in bacteria is getting worse and poses a serious problem for human health. The first factor that is contributing to this resistance is that antibiotics are widespread in the environment. Antibiotics are commonly used to treat infectious diseases in animals and so can sometimes be found in soil and water. Antibiotics are also being found in human foods because some animals are fed antibiotics to protect them from disease and help them grow faster. Antibotics are also commonly used in hospitals. The widespread use of antibiotics increases the exposure of bacteria to natural selection by antibiotics.

Additional factors that contribute to the development of bacterial resistance include:

- using antibiotics when they are not needed (such as for a common cold)
- using incorrect dosing amounts of antibiotics
- not finishing an entire course of antibiotic tablets.

How bacteria survive antibiotics

The increasing level of bacterial resistance to antibiotics occurs through natural selection. Mutation is part of this process. Mutations in genes have given some individual bacteria the ability to withstand antibiotics, through means such as:

- being able to destroy or inactivate the antibiotic molecules
- altering the bacterial cell wall so the antibiotic molecules cannot attach
- making it harder for antibiotic molecules to pass through, or permeate, the cell membrane
- being able to pump antibiotic molecules out of their cells so they cannot build up.

Transfer of genes between bacteria

Bacteria can develop resistance genes by gene transfer between bacteria.

One bacterial cell can transfer genes to another. This gene transfer can even occur between different species of bacteria.

Resistant bacteria can pass genetic information to non-resistant bacteria. The resistant bacterial cell forms a tube that joins the two bacterial cells and the resistant DNA passes through the tube to the other bacteria. This process is shown in Figure 4.4.1. Now both bacteria are resistant to an antibiotic that only one of them has been exposed to.

Many of the genes that give bacteria antibiotic resistance are found on plasmids. Plasmids are circular strands of DNA that are not part of the chromosome of the bacteria. Plasmids exist separately in the cell. It is not known how plasmids came to be in bacteria. Common bacteria now have combinations of genes that enable resistance to multiple antibiotics. This is largely due to plasmid transfer.

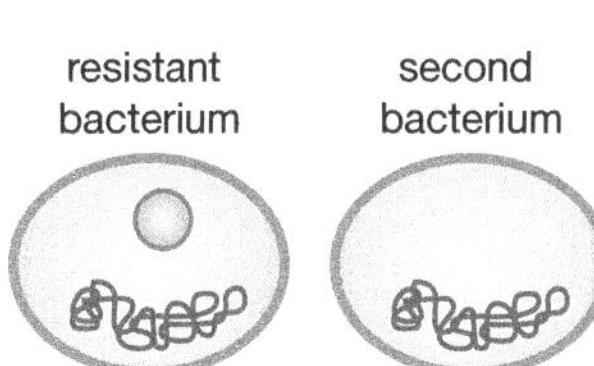

a The plasmid within the resistant bacterium contains resistant DNA.

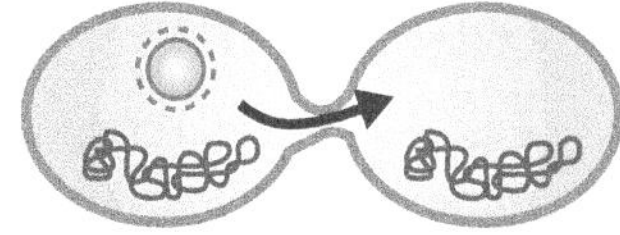

b A copy of the plasmid is carried into the second bacterium through a connecting tube.

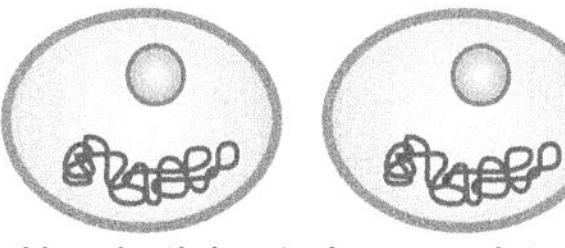

c Now both bacteria are resistant.

Figure 4.4.1 Gene transfer in bacteria

4.4 Combating bacterial resistance

1 List places where bacteria are being exposed to antibiotics.

2 List five factors that contribute to the development of resistance in bacteria.

3 Use the flow chart to show how antibiotics come to be found in human foods.

4 Suggest five ways in which antibiotics could enter the soil and water on farms.

5 List four characteristics resulting from mutation that have helped bacteria survive antibiotic molecules.

6 List two ways in which bacteria that have never been exposed to an antibiotic could contain genes that give them resistance to antibiotics.

7 Why does the transfer of genes between bacteria have the potential to be a serious health risk?

RATE MY UNDERSTANDING
Shade the face that shows your rating

4.5 Comparing eyes

Science understanding

FOUNDATION | **STANDARD** | ADVANCED

Humans have poor night vision. However, many animals such as cats have very good night vision. Most of the features that make night vision possible are found in the structure of the eye.

diurnal (*adj*) active during the day

nocturnal (*adj*) active at night

receptor (*n*) a group of cells that receive information

retina (*n*) a nerve layer at back of eye that senses light and receives the image

Structure of the retina

The retinas of diurnal and nocturnal animals have two major differences.

A The first difference is in the proportions of special light receptors called rods and cones. Cones sense colour but they do not detect enough light at night to send messages to the brain. This is why at night you cannot see the colours of objects. Rods are long, thin cells that work well in dim light.

Most nocturnal animals have many more rods in their retina than diurnal animals. So the eyes of nocturnal animals detect a lot more light at night. Many nocturnal animals, especially bats and lizards, have no cones at all. These nocturnal animals have excellent vision at night. Figure 4.5.1 shows a typical mammal eye shape (a cat) and typical structures of the retinas for a diurnal and a nocturnal animal.

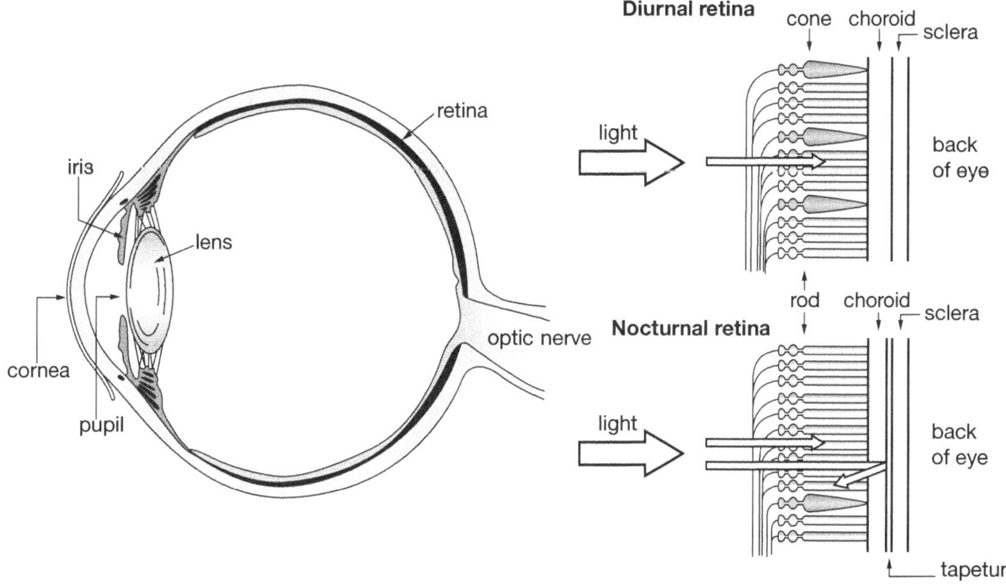

Figure 4.5.1 The retinas of diurnal and nocturnal animals

B The second difference is that the retina of some nocturnal animals has a layer of tissue called the tapetum. You can see its location in Figure 4.5.1. Light that has passed through the layer of rods and cones can strike the shiny surface of the tapetum and reflect back to the rods. This enables the rods to detect light that was missed the first time. This means that more light is collected, making it easier for the animal to see.

Eye size

Some nocturnal mammals, such as tarsiers, do not have a tapetum. The tarsier is a small primate of the Philippines that jumps through the trees, catching insects. It has the largest eyes of any animal in proportion to its body size. Figure 4.5.2 shows the incredible amount of space the eyes take up in the skull. Each eye is larger than the animal's brain.

4.5 Comparing eyes

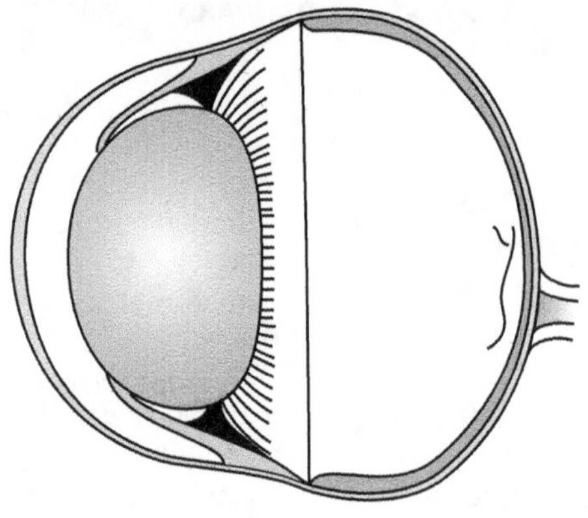

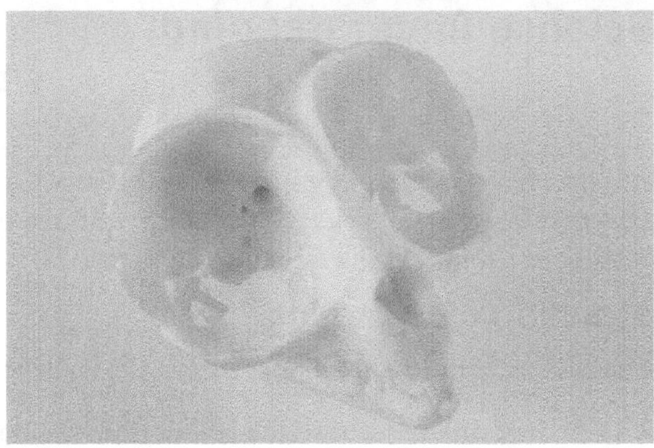

Figure 4.5.2 The eye and skull of a tarsier

Because tarsier eyes lack a tapetum, natural selection has favoured large eyes that can collect enough light for the rods in its retina to give clear vision. The tarsier eye is not spherical, but has become slightly tube shaped or elongated. The front of the eye has become enlarged to enable tarsiers to collect the maximum amount of light. The pupil and lens are huge, much bigger than in human eyes in proportion to the size of the eyeball. If the eye of a tarsier was spherical like a human eye, it would not fit in its skull!

pupil (*n*) the part of the eye that lets in light

spherical (*adj*) round like a ball

1 How does a tapetum increase the amount of light collected by the rods?

____

____

2 Compare the shape of a tarsier eye with the shape of a cat eye shown in Figure 4.5.1.

____

____

____

3 Explain how the lack of a tapetum in a tarsier is linked to the shape and size of its eye.

____

____

____

4 Use the flow chart to construct a likely series of events, explaining how natural selection may have changed the eye shape of the earliest ancestors of tarsiers as they became nocturnal animals.

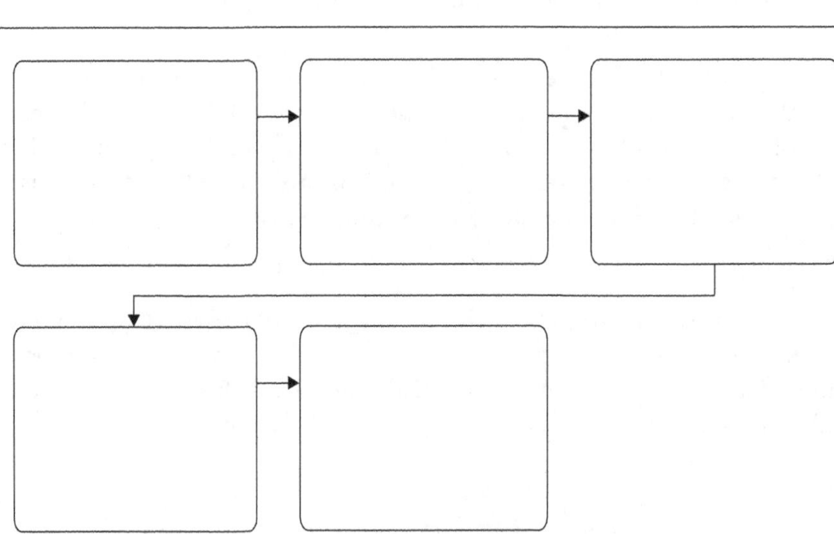

RATE MY UNDERSTANDING
Shade the face that shows your rating

4.6 The great tree of life

Science inquiry skills

FOUNDATION | **STANDARD** | ADVANCED

Processing & Analysing | Communicating

In his book *On the Origin of Species*, Charles Darwin drew a diagram he called the 'Great Tree of Life'. The 'tree' diagram showed how species should be related to each other if this theory of evolution was correct. Each branching of the tree showed where new species evolved because new structures appeared in the fossil record of the descendants.

phylogeny (*n*) a diagram showing the evolutionary relationships between species

Phylogeny

Darwin's technique is still used today. Often hundreds of features are used to create the 'tree'. Most of these are based on anatomy. However, now biologists also have molecular evidence such as DNA structure to help them. Consider Figure 4.6.1 which shows how tetrapods may have evolved. This diagram shows the branching tree structure originally developed by Darwin. This type of diagram is now called a phylogeny.

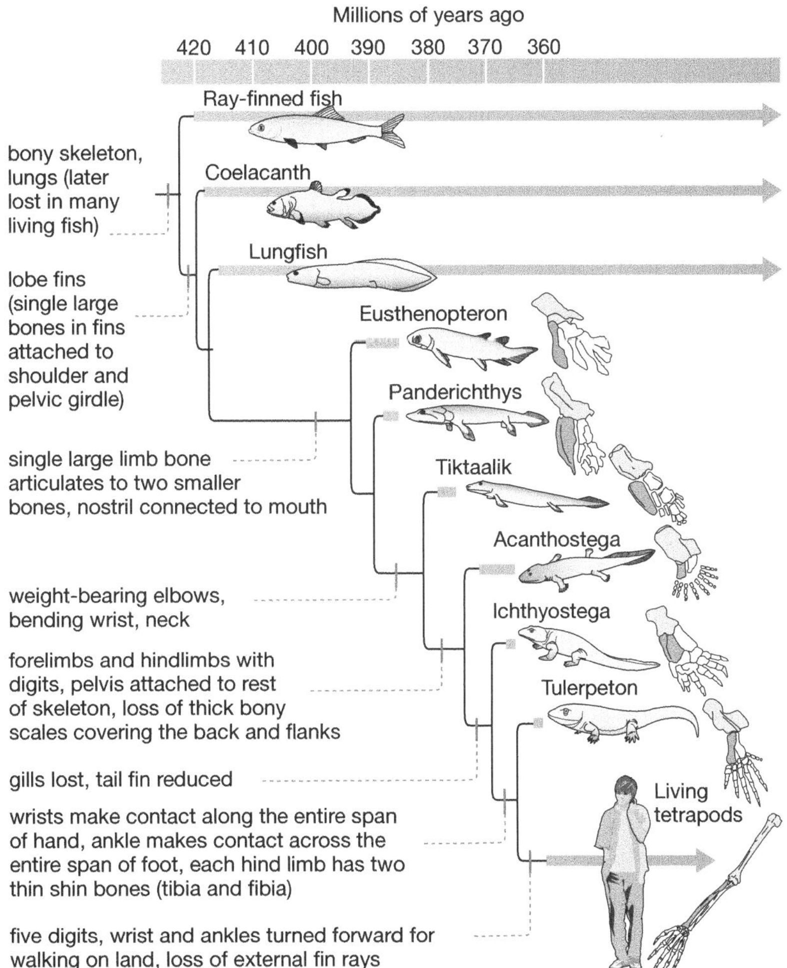

Figure 4.6.1 Phylogeny of tetrapods

Reading a phylogeny diagram

To read the diagram, start at the far left-hand side where the label says 'Bony skeleton, lungs'. You proceed through the diagram like you would use a biological key. The first line branches into two choices: 'ray-finned fish' and another branch that is labelled (with a small vertical line) 'lobe fins'. What the diagram shows is that the ray finned fish and the lobe-finned fish had a common ancestor. That ancestor had a bony skeleton and structures similar to lungs.

4.6 The great tree of life

The ancestor split into fish that evolved into both ray finned fish and lobe-finned fish. The lobe-finned fish eventually split into the coelacanths and another group that then further split. This group evolved into lungfish and another group which evolved limb bones having a humerus, radius and ulna.

Using a phylogeny like this, biologists are able to deduce which species should be the most closely related to each other. For example, consider which two of the following are the most closely related—ray finned fish, coelacanth or lungfish. The answer is the coelacanth and the lungfish. This is because at one point in time (about 425 million years ago based on the timeline at the top of the diagram) there were no coelacanths or lungfish, only a common ancestor of both. This common ancestor was a lobe-finned fish that lived at the same time as ray finned fish.

1 Explain what Darwin's 'Great Tree of Life' diagram showed.

____

____

2 Identify a new type of evidence used by modern biologists to create phylogenies.

____

____

(3) **(a)** Which two of the following are most closely related: *Eusthenopteron*, *Panderichthys* and *Tiktaalik*?

____

(b) Justify your answer.

____

____

(4) **(a)** Which two of the following are most closely related: lungfish, *Tiktaalik*, *Ichthyostega* and *Tulerpeton*?

____

(b) Justify your answer.

____

____

(5) Describe some features of the most recent ancestor of *Acanthostega* and *Ichthyostega*.

____

____

(6) Describe some features that were different between *Acanthostega* and *Ichthyostega*.

____

____

RATE MY UNDERSTANDING
Shade the face that shows your rating

4.7 Evolution of icefish

Science understanding

FOUNDATION | **STANDARD** | ADVANCED

Fish of the Family Channichthyidae, found in the waters around Antarctica, are very unusual. They have evolved in extremely cold water and are commonly called icefish (Figure 4.7.1). There are 25 species of icefish. Scientists have been interested in these fish since they were first discovered, because of their amazing ability to avoid freezing to death in extremely cold waters.

Icefish 'antifreeze'

Research has shown that an icefish's body contains special 'antifreeze' chemicals called glycoproteins, which seem to work by preventing ice crystals growing. The antifreeze chemical coats ice crystals so water cannot come into contact with the surface of the crystal and add to the size of the crystal. These chemicals are so effective that they are being researched for possible applications where freezing needs to be prevented. Possible applications include increasing the ability of crop plants to withstand frost, lengthening the shelf life of frozen foods, and potential applications in surgery, storing organs for transplant and treating hypothermia.

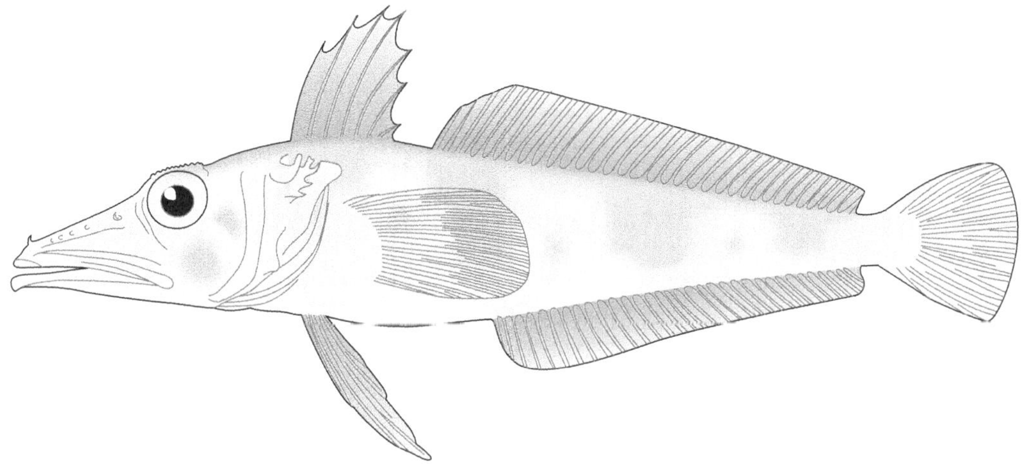

Figure 4.7.1 Ocellated icefish, *Chionodraco rastrospinosus*

Icefish blood

Another amazing fact about the icefish family is that they have almost no red blood cells. Red blood cells contain a chemical called haemoglobin, which bonds to oxygen. The haemoglobin transports oxygen around the bodies of all the other vertebrates. But icefish contain no haemoglobin and their blood is almost transparent. Oxygen is not very soluble so scientists wondered how oxygen was carried in the blood of these fish, and how their blood can carry enough oxygen to the cells to maintain respiration.

Research discovered that:

- the antifreeze production and the haemoglobin loss were both shown to be specific genetic mutations
- icefish have large hearts that pump out a high volume of blood each beat
- icefish have a very high volume of blood in their body
- icy cold water holds a lot more oxygen than warm water.

4.7 Evolution of icefish

Evolution and icefish

These icefish seemed to present an evolutionary puzzle. Scientists wondered why fish without red blood cells were not removed by natural selection.

The conclusion of scientists studying these fish is that they evolved in the following way:

- a mutation in an ancestral species allowed formation of the antifreeze chemicals
- a much later second mutation occurred which caused a loss of haemoglobin. This would normally be fatal in any vertebrate organism.
- the ancestors of the icefish were present in Antarctica as the climate began cooling
- the loss of haemoglobin was not fatal as colder water could carry more oxygen than haemoglobin and the antifreeze helped them cope with the cold water
- there would not be much competition from other fish since those fish lacked the antifreeze chemicals and most seem to have been unable to survive the cold.

1 What adaptation of icefish enables them to maintain liquid water in their bodies despite the extreme cold?

____

____

2 Describe three adaptations of icefish that enabled them to survive without haemoglobin.

____

____

____

(3) Explain how natural selection could have acted on all Antarctic fish and selected in favour of icefish.

____

____

____

[4] Why does it make sense that the antifreeze mutation came before the haemoglobin mutation?

____

____

____

[5] Propose whether the adaptations in the circulatory system of icefish came before or after the loss of haemoglobin.

____

____

____

RATE MY UNDERSTANDING
Shade the face that shows your rating

4.8 Bipedalism

Science inquiry skills

FOUNDATION | **STANDARD** | ADVANCED

The main feature of the sub-tribe of animals to which modern humans belong is that they walk upright on two legs. This feature is called bipedalism. A summary of the adaptations that humans have for upright walking are:

- an S-shaped spine with lumbar curve and wedge-shaped vertebrae—this bends the vertebral column into an upright position instead of sloping forward as in chimps and gorillas (see diagrams 1 and 2 below)
- a wide, short pelvis—this allows attachment of muscles between the legs and pelvis, which support the upper body as we walk (see Diagram 1 below, and Diagram 5 on Page 66)
- large hip sockets and long legs—this enables large legs that support weight (see Diagram 3)
- carrying angle of femur—knees bent inwards for more stable standing and walking (see diagrams 3 and 7)
- a femur with enlarged outer condyle—this bears much of the weight in the knee (see Diagram 7)
- arches in the foot—this allows for energy absorption and storage while walking
- an enlarged first metatarsal and big toe—this allows for weight support while walking (see Diagram 4)
- a central foramen magnum (the place where the spinal cord exits the skull)—this allows the skull to balance on top of the vertebral column without large muscle support (see diagrams 1 and 6).

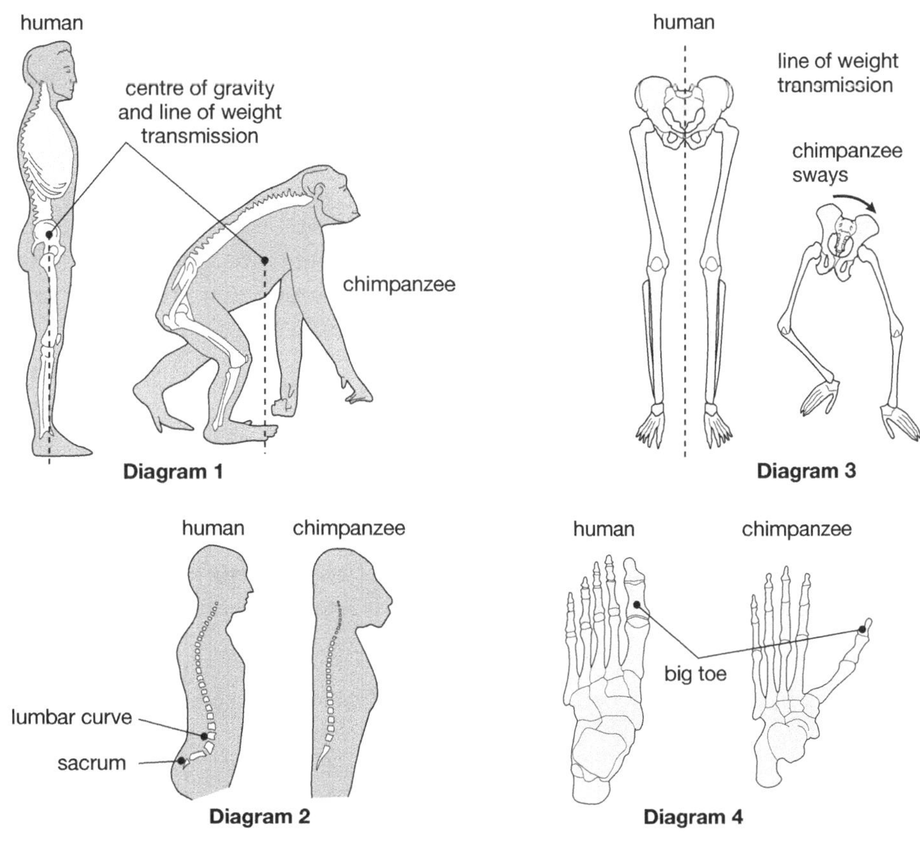

4.8 Bipedalism

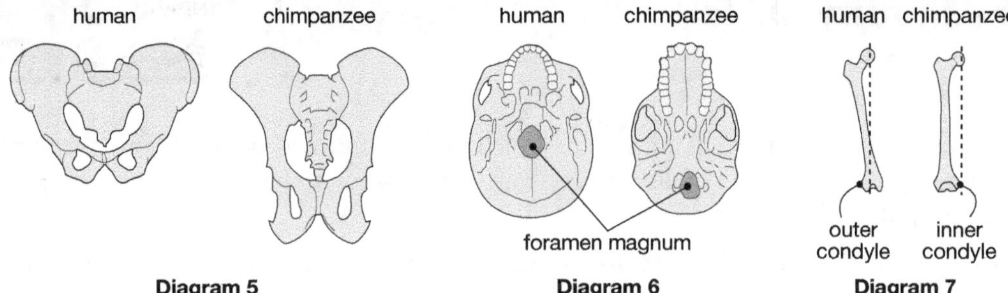

Diagram 5

Diagram 6

Diagram 7

To understand how humans are adapted for bipedalism, try these activities.

1 Stand with your shoulder to the wall. Stand straight up with your hands down by your sides. Now slowly lift the leg furthest from the wall sideways away from you.

Describe what happens to you.

____

____

2 Stand away from the wall and do the same thing as in question 1, but try to stay standing. Feel what your head does as you lift your leg.

Why did you fall in the first situation when standing against the wall, but not in the second when away from the wall?

____

____

3 In Diagram 3, the dotted line shows the line down which body weight can be considered to act (through the centre of gravity). The knees of a human are positioned close to this line. In a chimpanzee, the knees are much further apart and not as close to the line of weight transmission. When a chimp walks, the only way it can get its body weight to be supported by a leg is to tilt its body sideways towards the leg on the ground. Otherwise, the weight would not pass down through the leg on the ground and it would fall.

How does this compare with the activity you tried in question 2?

____

____

4 You do not sway as much as chimps when you walk because your knees bring your foot near the centre of your body. Try walking along a straight line and watch where your feet go and you will see how it works.

What did you observe about where your feet landed as you walked?

____

____

5 Suggest why the carrying angle of the femur is an adaptation for upright walking.

____

____

RATE MY UNDERSTANDING
Shade the face that shows your rating   

4.9 Literacy review

Science understanding

FOUNDATION	STANDARD	ADVANCED

1 The table below contains some terms related to natural selection and evolution. Complete the table by writing a definition for each term.

Term	Definition
evolution	
natural selection	
species	
generation	
primates	
embryology	
selective agent	

2 Write the correct term for each definition in the table below.

Definition	Term
selective breeding between two closely related individuals	
using technology to gather, store and analyse biological data and to solve biological questions	
the effect of the selective agent on a population	
the physical structure of an organism	
a group or species that evolves from an ancestor	
a rapid variation or divergence of an evolutionary lineage from a recent common ancestor	
the term for structures in different species that are controlled by some of the same inherited genes	
differences in characteristics due to differences in particular genes	
a genus to which humans and several extinct species belong	

RATE MY UNDERSTANDING
Shade the face that shows your rating

4.10 Thinking about my learning

Think carefully about what you have learned about the processes which lead to species changing over time. Also, think carefully about how you approached this unit of work. Then complete the following sentences.

While learning about evolution and natural selection I was surprised to find out that ...
The most interesting thing I learned was ...
One concept I found very easy to understand was ...
One idea I found quite challenging was ...
The strategy I used to help me to understand this challenging concept was ...
I am still not confident that I fully understand ...
In order to help me understand this better I am going to ...
While studying this topic the activity I enjoyed the most was ...
When studying the next topic some things I will do differently are ...

CHAPTER 5

The periodic table

5.1 Knowledge preview

FOUNDATION | STANDARD | ADVANCED

(1) In the space below draw a simple labelled diagram to show what you know about the parts of an atom.

(2) List the names and symbols for 10 common elements in the periodic table.

_______________ _______________ _______________ _______________ _______________

_______________ _______________ _______________ _______________ _______________

(3) Define the following terms:

(a) atom ____

(b) element ____

(c) compound ____

(d) crystal lattice. ____

(4) **(a)** Compare an atom and an ion, listing their similarities and differences.

____

____

____

(b) What is the difference between a positive ion (cation) and a negative ion (anion)? In your answer, include how each ion forms.

____

____

____

5.1 Knowledge preview

5. Fill in the table below regarding metals and non-metals.

	Metals	Non-metals
Examples of elements		
Properties		
Description of their location on the periodic table		

6. What is a row called on the periodic table and how many rows are there?

7. Circle the correct answer for each statement.

(a) Elements are arranged on the periodic table according to this.

A mass

B atomic number

C number of electrons

(b) These are the symbols for the elements oxygen and sodium.

A O and SO

B O and Na

C Ox and Na

(c) Group 1 elements have this property.

A react with water

B are metals

C have a single electron in their outer shell

(d) The chemical properties of an element are mainly influenced by this.

A number of electrons

B number of neutrons

C number of protons

5.2 Periodic nature of the elements

Science inquiry skills

FOUNDATION | STANDARD | ADVANCED

Processing & Analysing | Communicating

Table 5.2.1 shows the melting points, boiling points and atomic radii of the first 20 elements of the periodic table.

Table 5.2.1 Melting points, boiling points, and atomic radii of elements table

Element	Atomic number	Melting point (°C)	Boiling point (°C)	Atomic radius (Å)
hydrogen, H	1	–259	–253	2.1
helium, He	2	–272	–269	1.9
lithium, Li	3	181	1347	1.6
beryllium, Be	4	1278	2970	1.1
boron, B	5	2300	2550	1.0
carbon, C	6	3500	4827	0.9
nitrogen, N	7	–210	–196	0.9
oxygen, O	8	–218	–183	0.7
fluorine, F	9	–220	–188	0.6
neon, Ne	10	–249	–246	0.5
sodium, Na	11	98	883	1.9
magnesium, Mg	12	650	1107	1.6
aluminium, Al	13	660	2467	1.4
silicon, Si	14	1410	2355	1.3
phosphorus, P	15	44	280	1.3
sulfur, S	16	113	445	1.3
chlorine, Cl	17	–101	-35	1.0
argon, Ar	18	–189	–186	0.9
potassium, K	19	64	774	2.4
calcium, Ca	20	839	1487	2.0

1. Construct a line graph showing how boiling point varies with increasing atomic number. Join your points up dot-to-dot. Use the grid and the vertical axis on the left of the grid on page 72 to help you.

2. On the same grid construct another line graph showing how melting point varies with atomic number. Once again, use the same vertical axis on the left.

3. Use the vertical axis on the right of the grid to construct another line graph showing how the atomic radius varies with atomic number.

4. Compare the overall shape of the melting and boiling point graphs.

5. Describe the overall trend in the melting point, boiling point and atomic radius as you move across a period of the periodic table.

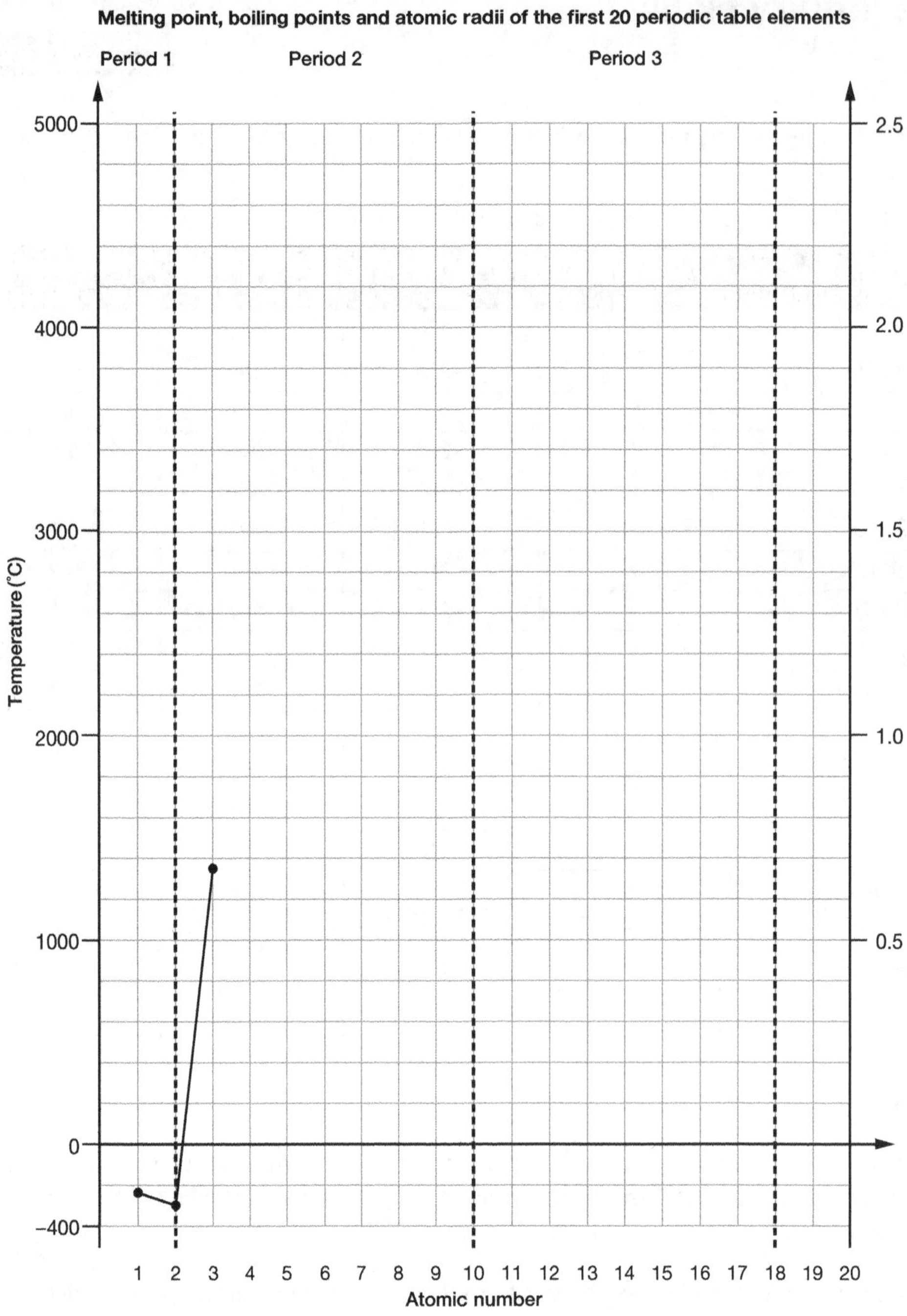

Melting point, boiling points and atomic radii of the first 20 periodic table elements

Period 1 Period 2 Period 3

Temperature (°C): 5000, 4000, 3000, 2000, 1000, 0, –400

Atomic radius (Å): 2.5, 2.0, 1.5, 1.0, 0.5

1 2 3 4 5 6 7 8 9 10 11 12 13 14 15 16 17 18 19 20

Atomic number

(6) Describe what happens to the trend in the melting point, boiling point and atomic radius when you move from one period to another.

____

____

(7) Use your graph to determine what happens to the size of an atom as you move down a group of the periodic table (for example, moving down group 1 from H to Li to Na to K.

____

____

RATE MY UNDERSTANDING
Shade the face that shows your rating

5.3 Scientists of the periodic table

Science as a human endeavour

FOUNDATION | **STANDARD** | ADVANCED

Many scientists contributed to our understanding of the elements, their properties and the family relationships between them. Therefore each of these scientists directly or indirectly contributed to the development of the periodic table we use today.

Scientist	Lived	Scientific contributions	Other facts
Antoine-Laurent de Lavoisier	1743–93	• identified and named the existing 33 elements known at the time • proved that mass was not destroyed in a reaction • proved that rusting involved oxygen bonding to iron (increasing its mass) • proved that combustion needed oxygen	• nationality: French • was incredibly wealthy, working as a tax collector • had 13 000 beakers in his laboratory • was executed by guillotine after the French Revolution • his statue in Paris was made with a 'spare' head from another statue • believed that light was an element
John Dalton	1766–1844	• revived the Ancient Greek idea of atoms • gave each element a pictorial symbol, which is not used today • predicted atomic masses • developed laws explaining how gases behaved • showed that compounds were made of elements in definite mass proportions	• nationality: English • came from a very poor family • was put in charge of his small school when he was 12 years old • still taught mathematics at a primary school when elderly and famous
Jons Jacob Berzelius	1779–1848	• used atomic mass to arrange elements • gave elements symbols that we still use today based on their Greek or Latin names • showed that compounds were made of elements in definite mass proportions • was the first to classify substances as organic (containing C) or inorganic (containing no C)	• nationality: Swedish • used superscript numbers instead of subscript numbers in chemical formulas—wrote water as H^2O instead of H_2O
Johann Dobereiner	1780–1849	• placed elements into 'triads', families of three	• nationality: German
John Newlands	1837–98	• used atomic mass to arrange elements • noticed every eighth element 'harmonising' (forming patterns) • placed elements in periods of 7 elements	• nationality: English • called his arrangement 'the law of octaves' after musical octaves
Dmitri Ivanovich Mendeleev	1834–1907	• used atomic mass to arrange elements • like Newlands, arranged the elements in periods of 7 • arranged related elements in columns like Dobereiner's triads • left gaps for undiscovered elements (e.g. Ge)	• nationality: Russian • refused to accept that electrons, and radiation and radioactivity existed • has a radioactive element named after him—mendelevium!

5.3 Scientists of the periodic table

Scientist	Lived	Scientific contributions	Other facts
Lothar Meyer	1830–95	• constructed a table very similar to Mendeleev's • finished his periodic table before Mendeleev • published his table after Mendeleev	• nationality: German
William Ramsay	1852–1916	• discovered the noble, or inert, gases • set them as a separate group	• nationality: Scottish • received the Nobel Prize in Chemistry
Henry Moseley	1887–1915	• used atomic number and not mass to arrange elements • produced the periodic table still in use today	• nationality: English • was shot through the head and killed in World War I at Gallipoli

1 Use the information in the table and further research to identify the scientist who:

(a) was Scottish ______________________

(b) finished his table before Mendeleev but received less recognition than him

____

(c) thought the elements harmonised in octaves like music ______________________

(d) received a Nobel Prize ______________________

(e) wrote water as H^2O ______________________

(f) left gaps in his table ______________________

(g) developed triads ______________________

(h) thought that light was an element ______________________

(i) developed gas laws ______________________

(j) arranged elements according to their atomic number and not mass ______________________

(k) was shot in the head ______________________

(l) had his head chopped off ______________________

(m) was Russian ______________________

(n) revived the idea that atoms existed ______________________

(o) had 13 000 beakers in his laboratory ______________________

(p) was put in charge of his school when he was 12 ______________________

(q) grouped chemicals as organic or inorganic ______________________

(r) gave the elements known at that time the symbols we use today ______________________

(s) has a radioactive element named after him, even though he didn't believe in radioactivity

____

(t) proved that iron got heavier when it rusted. ______________________

RATE MY UNDERSTANDING
Shade the face that shows your rating

5.4 Families of the periodic table

Science understanding

FOUNDATION	STANDARD	ADVANCED

KEY

- non-metals
- metals
- metalloids

atomic number — 13
symbol — **Al**
name — aluminium

1	2	3	4	5	6	7	8	9	10	11	12	13	14	15	16	17	18
1 **H** hydrogen																	2 **He** helium
3 **Li** lithium	4 **Be** beryllium											5 **B** boron	6 **C** carbon	7 **N** nitrogen	8 **O** oxygen	9 **F** fluorine	10 **Ne** neon
11 **Na** sodium	12 **Mg** magnesium											13 **Al** aluminium	14 **Si** silicon	15 **P** phosphorus	16 **S** sulfur	17 **Cl** chlorine	18 **Ar** argon
19 **K** potassium	20 **Ca** calcium	21 **Sc** scandium	22 **Ti** titanium	23 **V** vanadium	24 **Cr** chromium	25 **Mn** manganese	26 **Fe** iron	27 **Co** cobalt	28 **Ni** nickel	29 **Cu** copper	30 **Zn** zinc	31 **Ga** gallium	32 **Ge** germanium	33 **As** arsenic	34 **Se** selenium	35 **Br** bromine	36 **Kr** krypton
37 **Rb** rubidium	38 **Sr** strontium	39 **Y** yttrium	40 **Zr** zirconium	41 **Nb** niobium	42 **Mo** molybdenum	43 **Tc** technetium	44 **Ru** ruthenium	45 **Rh** rhodium	46 **Pd** palladium	47 **Ag** silver	48 **Cd** cadmium	49 **In** indium	50 **Sn** tin	51 **Sb** antimony	52 **Te** tellurium	53 **I** iodine	54 **Xe** xenon
55 **Cs** caesium	56 **Ba** barium	57–71 lanthanoids	72 **Hf** hafnium	73 **Ta** tantalum	74 **W** tungsten	75 **Re** rhenium	76 **Os** osmium	77 **Ir** iridium	78 **Pt** platinum	79 **Au** gold	80 **Hg** mercury	81 **Tl** thallium	182 **Pb** lead	83 **Bi** bismuth	84 **Po** polonium	85 **At** astatine	86 **Rn** radon
87 **Fr** franchium	88 **Ra** radium	89–103 actinoids	104 **Rf** rutherfordium	105 **Db** dubnium	106 **Sg** seaborgium	107 **Bh** bohrium	108 **Hs** hassium	109 **Mt** meitnerium	110 **Ds** darmstadium	111 **Rg** roentgenium	112 **Cn** copernicium	113 **Uut** ununtrium	114 **Fl** flerovium	115 **Uup** ununpentium	116 **Lv** livermorium	117 **Uus** ununseptium	118 **Uuo** ununoctium

lanthanides	57 **La** lanthanum	58 **Ce** cerium	59 **Pr** praseodymium	60 **Nd** neodymium	61 **Pm** promethium	62 **Sm** samarium	63 **Eu** europium	64 **Gd** gadolinium	65 **Tb** trebium	66 **Dy** dysprosium	67 **Ho** holmium	68 **Er** erbium	69 **Tm** thulium	70 **Yb** ytterbium	71 **Lu** lutetium
actinides	89 **Ac** actinium	90 **Th** thorium	91 **Pa** protactinium	92 **U** uranium	93 **Np** neptunium	94 **Pu** plutonium	95 **Am** americum	96 **Cm** curium	97 **Bk** berkelium	98 **Cf** californium	99 **Es** einsteinium	100 **Fm** fremium	101 **Md** mendelevium	102 **No** nobelium	103 **Lr** lawrencium

1 Use different colours or shading to demonstrate where the following elements are on the simple version of the periodic table shown above. Use the box next to each group of elements as a key to explain the colouring on the periodic table. Note that some of your coloured or shaded areas might overlap.

(a) ☐ noble or inert gases

(b) ☐ alkali metals

(c) ☐ actinides

(d) ☐ transition elements

(e) ☐ halogens

(f) ☐ alkaline earths

(g) ☐ lanthanides

2 Identify the element described below by placing a:

(a) circle around the symbol of the only non-metal that exists as a liquid at normal room temperatures

(b) square around the symbol of the element found in all living (and dead) things, being the basis for most of their molecules

(c) triangle around the symbol of the lightest of all elements

(d) oval around the symbol of the heaviest of all elements.

RATE MY UNDERSTANDING Shade the face that shows your rating	☺	😐	☹

5.5 Organic chemistry

Science understanding

FOUNDATION | STANDARD | **ADVANCED**

A carbon atom has four outer-shell electrons, placing it in group 14 (or group IV) of the periodic table.

When it forms a molecule, these four electrons are shared with other carbon atoms or atoms of other non-metals to form four bonds. This ability to form four bonds gives carbon the capacity to form some amazing molecules.

Carbon-based molecules range from small groupings of atoms to long chains and rings that have carbon atoms as their backbone. Some are shown in Figure 5.5.1 Each bond is shown as a single 'stick' between the atoms. If there are two 'sticks' then there is a double bond (two single bonds) holding the atoms together.

These carbon backbones give the molecules a new classification too. They are classified as organic. Organic chemistry is the study of these carbon molecules, their properties and their reactions.

Organic molecules form the basis of life on Earth. They make up the:

- living tissue of animals, such as their skin and organs
- leaves, stems, trunks and roots of plants
- cell walls of bacteria and other micro-organisms.

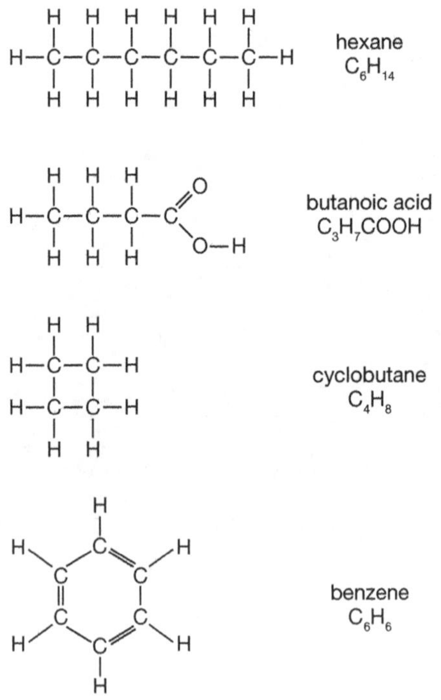

hexane C_6H_{14}

butanoic acid C_3H_7COOH

cyclobutane C_4H_8

benzene C_6H_6

Figure 5.5.1

Carbon becomes part of plants via photosynthesis, in the following reaction:

$6CO_2$	+	$6H_2O$	→	$C_6H_{12}O_6$	+	$6O_2$
carbon dioxide		water		glucose		oxygen gas

During photosynthesis plants take in carbon dioxide, break it up and construct a new organic molecule called glucose ($C_6H_{12}O_6$). Some of this is used immediately in respiration to give the energy the plant needs. The respiration reaction is:

$C_6H_{12}O_6$	+	$6O_2$	→	$6CO_2$	+	$6H_2O$
glucose		oxygen gas		carbon dioxide		water

The rest of the glucose is stored within the plants in the form of other organic molecules such as sucrose ($C_{12}H_{22}O_{11}$) and starch. Animals eat plants or eat other animals that eat plants. In this way, the carbon stored in the plant becomes part of the animals' structures too.

Dead organisms also contain organic molecules. For this reason, substances such as wood and paper (from plants) and leather (from animals) are organic. Fossil fuels such as oil, petrol, coal and gas are organic too because they are made from the remains of long-dead organisms. Plastics are made from oil and so they are organic as well.

When burnt, wood, paper, plastics and fossil fuels release much of their carbon into the atmosphere as carbon dioxide (CO_2). For example, petrol contains octane (C_8H_{18}).

Its combustion reaction is:

$2C_8H_{18}$	+	$25O_2$	→	$16CO_2$	+	$18H_2O$
octane		oxygen		carbon dioxide		water

5.5 Organic chemistry

Carbon dioxide is a known greenhouse gas. Studies have shown that its increasing concentration in the atmosphere is causing global warming. Average global temperatures are rising as a result, leading to a change in the climate.

1 For the molecules in Figure 5.5.1, state how many bonds each carbon atom has.

____

(2) Vegetables, meat and other products are often described as organic if they are pesticide-free and herbicide-free and only fertilised with natural compost and manure. Suggest reasons why the term organic is used this way when it has such a different scientific meaning.

____

____

(3) Suggest what the following scientists study:

(a) organic chemist ______________________________

(b) inorganic chemist. ______________________________

(4) Compare the structures of hexane and benzene by listing their similarities and differences.

(a) similarities ______________________________

(b) differences ______________________________

(5) Explain where the carbon in the following comes from.

(a) plants ______________________________

(b) animals ______________________________

(c) fossil fuels ______________________________

(d) plastics ______________________________

(6) Explain why oil, petrol, gas and coal are known as fossil fuels.

____

____

(7) Explain why many scientists link climate change to the combustion of fossil fuels.

____

____

RATE MY UNDERSTANDING
Shade the face that shows your rating

5.6 Hydrocarbons

Science understanding

FOUNDATION | STANDARD | **ADVANCED**

There are many thousands of different organic molecules and they must be named in a systematic way to avoid confusion. The International Union of Pure and Applied Chemistry (IUPAC) has developed precise rules on how to name organic molecules.

The simplest of all organic molecules are the hydrocarbons. As their name implies, these molecules are made of just carbon and hydrogen atoms. Depending on their bonding, hydrocarbons can be classified as alkanes, alkenes or alkynes. The first ten of each family are shown in Table 5.6.1.

Table 5.6.1 Hydrocarbon family

	Alkanes	Alkenes	Alkynes
general formula	C_nH_{2n+2}	C_nH_{2n}	C_nH_{2n-2}
$n = 1$	methane, CH_4 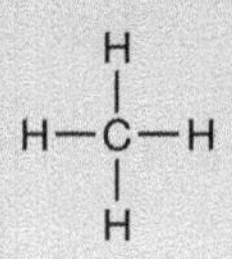	none exists	none exists
$n = 2$	ethane, C_2H_6 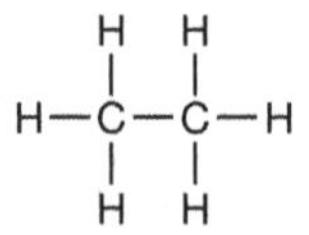	ethene, C_2H_4 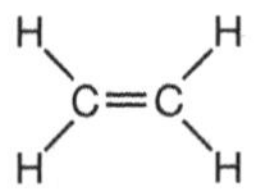	ethyne, C_2H_2 H—C≡C—H
$n = 3$	propane, C_3H_8 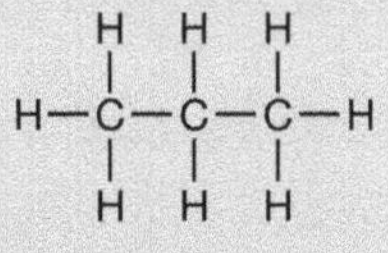	propene, C_3H_6 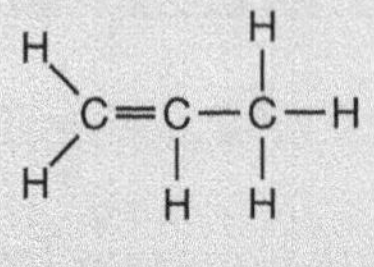	propyne, C_3H_4 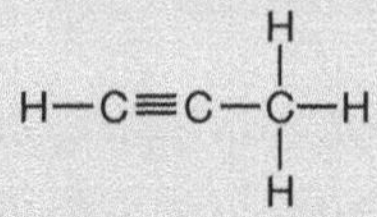
$n = 4$	butane, C_4H_{10}	butene, C_4H_8	butyne, C_4H_6
$n = 5$	pentane, C_5H_{12}	pentene, C_5H_{10}	pentyne, C_5H_8
$n = 6$	hexane, C_6H_{14}	hexene, C_6H_{12}	hexyne, C_6H_{10}
$n = 7$	heptane, C_7H_{16}	heptene, C_7H_{14}	heptyne, C_7H_{12}
$n = 8$	octane, C_8H_{18}	octene, C_8H_{16}	octyne, C_8H_{14}
$n = 9$	nonane, C_9H_{20}	nonene, C_9H_{18}	nonyne, C_9H_{16}
$n = 10$	decane, $C_{10}H_{22}$	decene, $C_{10}H_{20}$	decyne, $C_{10}H_{18}$

1. Name the organisation that determines the names of different organic compounds.

____

(2) Different scientists might speak different languages but they all use the same names for chemicals. What do you think science would be like if they didn't?

(3) Identify which family of hydrocarbons has:

(a) only single bonds

(b) a double bond

(c) a triple bond.

(4) Compare the alkanes, alkenes and alkynes by listing their similarities and differences.

(a) similarities

(b) differences

(5) Use the general formulas given in Table 5.6.1 to calculate the chemical formulas of the:

(a) 12th member of the alkanes ($n = 12$)

(b) 20th member of the alkenes ($n = 20$)

(c) 30th member of the alkynes ($n = 30$).

(6) To get the next member of the alkane family, you add a grouping of carbon and hydrogen atoms. Identify which of the following groupings needs to be added to get the next alkane: CH, CH_2, CH_4 or CH_6.

(7) Use the list in question 6 to identify the grouping that needs to be added to get the next:

(a) alkene

(b) alkyne.

8 Propose reasons why there is no alkene or alkyne with just one carbon atom.

5.6 Hydrocarbons

9 Use the naming rules for organic compounds to propose possible names of the following hydrocarbons.

(a)

```
    H       H   H
    |       |   |
H — C — C = C — C — H
    |   |       |
    H   H       H
```

____

(b)

```
    H   H           H   H
    |   |           |   |
H — C — C — C ≡ C — C — C — H
    |   |           |   |
    H   H           H   H
```

____

(c)

```
            H
            |
        H — C — H
  H         |         H
  |         |         |
H—C ——————— C ——————— C—H
  |         |         |
  H         |         H
        H — C — H
            |
            H
```

____

10 Construct diagrams of what you think the following molecules would look like.

(a) butene

(b) pentane

(c) octyne

RATE MY UNDERSTANDING
Shade the face that shows your rating

5.7 Electron configuration and shells

Science understanding

FOUNDATION | **STANDARD** | ADVANCED

Electrons are arranged in shells around the nucleus of an atom. Electrons fill the shell closest to the nucleus first, then the second, then the third shell and so on. A maximum of 2 electrons can fill the first shell ($n = 1$), 8 the second shell ($n = 2$), 18 the third shell ($n = 3$) and 32 the fourth shell ($n = 4$). The numbers of electrons that a shell can hold is calculated using the formula $2n^2$, where n is the number of the electron shell.

1. Use this information to show how the electrons are arranged in the first 18 elements of the periodic table. Sodium (Na) has already been done for you.

1 H________ **2** He________ 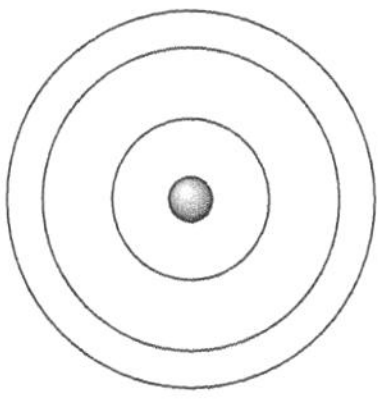 **3** Li________ 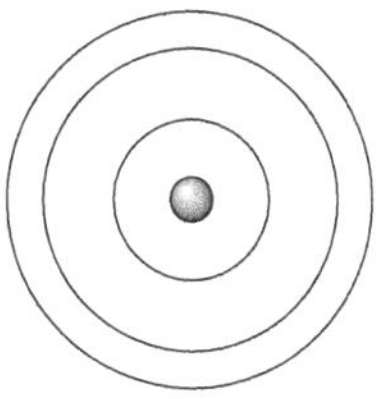 **4** Be________

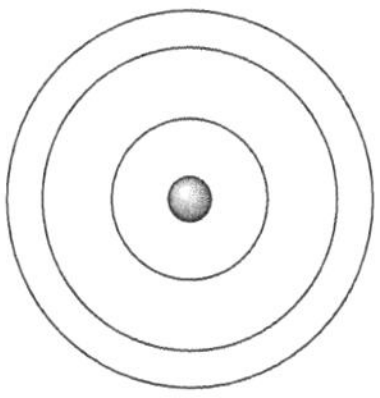

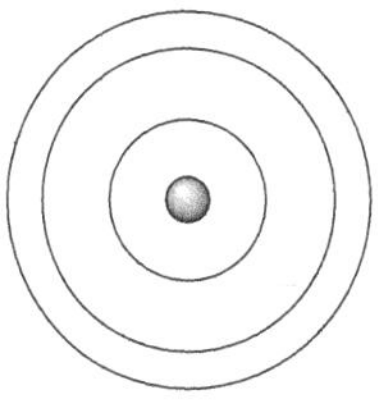

5 B________ 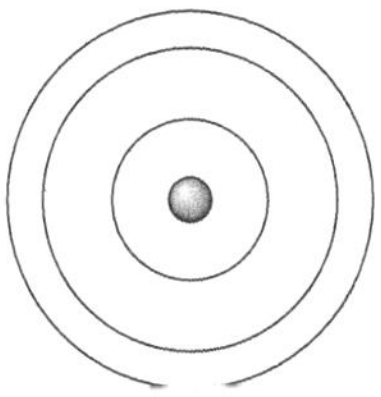 **6** C________ 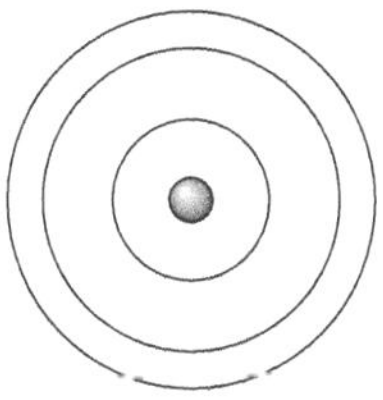 **7** N________ 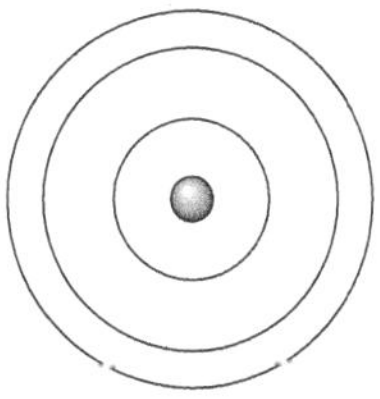 **8** O________

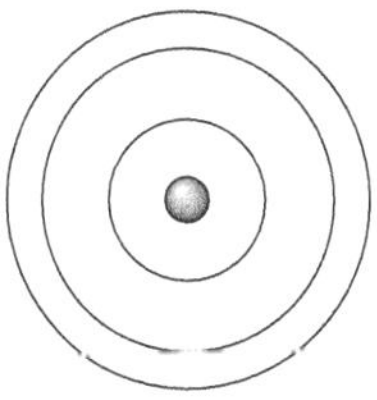

9 F________ 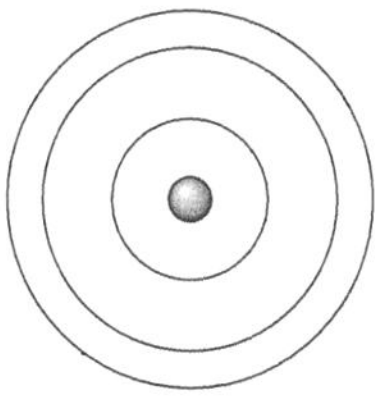 **10** Ne________ 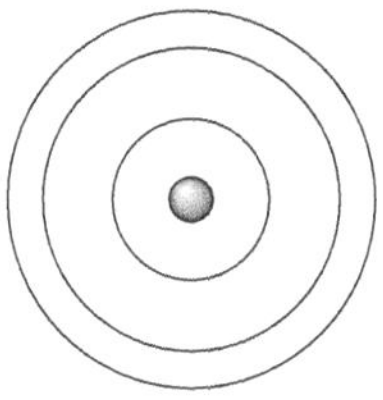 **11** Na 2,8,1

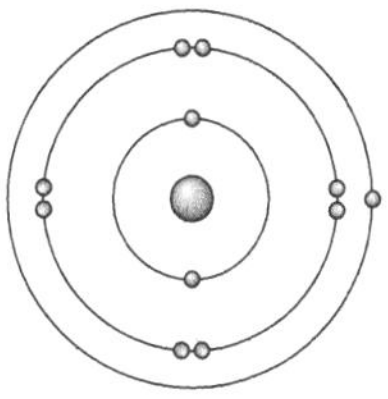

HINT

Electrons pair up once there are four electrons in a shell. Sodium has 8 electrons arranged in four pairs in its second shell.

12 Mg________ **13** Al________ 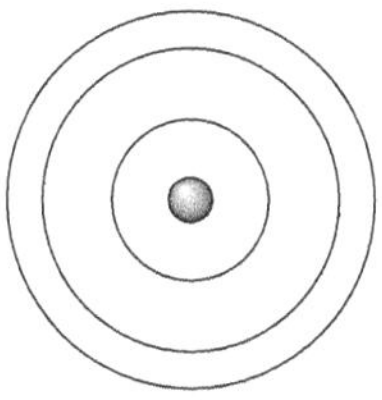 **14** Si________ 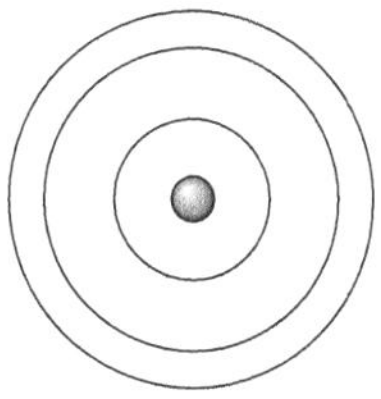 **15** P________

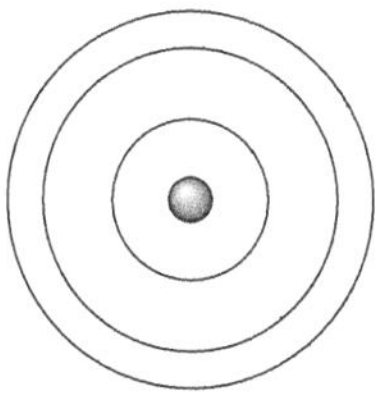

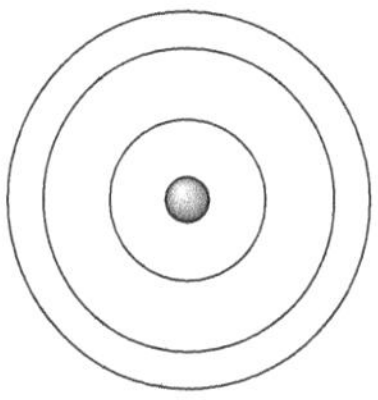

16 S________ 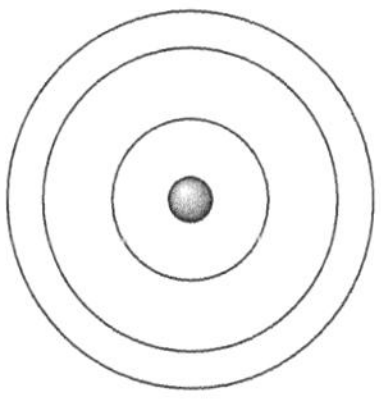 **17** Cl________ 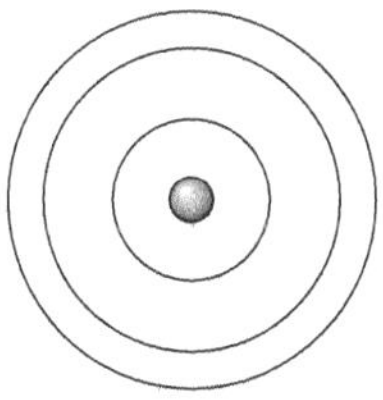 **18** Ar________

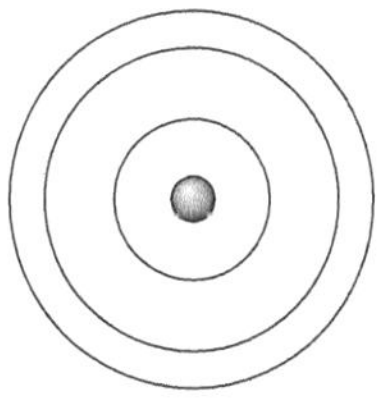

RATE MY UNDERSTANDING
Shade the face that shows your rating

5.8 Literacy review

Science understanding

FOUNDATION | STANDARD | ADVANCED

1 Recall your knowledge of the periodic table to complete this crossword puzzle.

ACROSS

4 Metals that are in group 1
6 A special block of elements in groups 3–12
8 Element with the symbol Na
9 One form of carbon that is incredibly hard
11 Scientist who left gaps in his periodic table for elements yet to be discovered
13 There are 118 of these known to exist
15 Non-metals that are in group 17
16 The lightest of all the elements
18 Vertical columns in the periodic table

DOWN

1 One of two special blocks of elements at the bottom of the periodic table
2 Another name for the noble gases
3 'Earths' that are in group 2
4 Different forms in which carbon can exist
5 Horizontal rows in the periodic table
7 Atoms of the same element with different numbers of neutrons
10 Number that tells you how many protons are in the nucleus
11 Arranged his periodic table according to atomic number and not atomic mass
12 Particles in the atom without any charge
14 Number showing the total number of particles in the nucleus
17 Group 18 gases

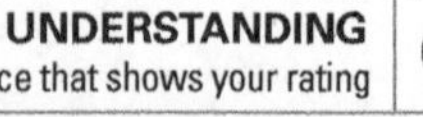

5.9 Thinking about my learning

1 Tick the square that best matches your understanding for each of the big ideas.

	Big ideas	I still need help with this	I understand this	I understand this well and can teach someone about it
Science understanding	I can describe the structure of an atom in terms of electron shells.			
	I understand that elements are the building blocks of all substances.			
	I can explain what atomic number and mass number are.			
	I can explain how the isotopes of an element differ.			
	I understand that the periodic table arranges elements by atomic number.			
	I can explain why elements grouped together on the periodic table have similar properties.			
	I can identify the periods on the periodic table and describe the properties of elements in them.			
	I can name the groups on the periodic table and describe the properties of elements in them.			
Science inquiry skills	I can follow a method and set up experiments involving properties of elements.			
	I can write experimental reports and discuss and explain my experiment results.			
	I can work safely in the laboratory.			
Science as a human endeavour	I understand that knowledge about elements and the periodic table has developed gradually from the contributions of many scientists.			

(a) Describe one thing that you found interesting.

____

____

(b) Describe one thing that you found difficult.

____

____

(c) Describe one question that you still have.

____

____

CHAPTER 6

Chemical reactions

6.1 Knowledge preview

Science understanding

FOUNDATION | STANDARD | ADVANCED

1 Fill in the table below with the names and formulas for each substance.

Name of substance	Chemical Formula
hydrogen gas	
water	
oxygen gas	
carbon dioxide	
hydrochloric acid	HCl

Name of substance	Chemical Formula
	$NaCl$
	$CaCO_3$
	H_2SO_4
	CH_4

2 Explain what the following terms mean:

(a) chemical reaction ______________________

(b) reactants ______________________

(c) products. ______________________

3 List the ions that would form from the following elements. The first one has been done for you:

Element	Name of ion that it forms	Symbol for ion
hydrogen	hydrogen ion	H^+
sodium		
chlorine		

Element	Name of ion that it forms	Symbol for ion
oxygen		
beryllium		
aluminium		

4 Name the following ions:

(a) NO_3^- ______________________

(b) OH^- ______________________

(c) SO_4^{2-} ______________________

6.2 Combustion of magnesium

Science inquiry skills

FOUNDATION | **STANDARD** | ADVANCED

Processing & Analysing **Communicating**

Before scientists understood the atomic theory of matter, people thought that the process of combustion involved the release of a material they called 'phlogiston'. Today, scientists know that phlogiston does not exist—combustion occurs when chemicals combine vigorously with oxygen.

failings (*n*) fault, problem
vigorously (*adj*) with a lot of energy

One of the failings of the phlogiston theory was that it could not explain the combustion of metals. The phlogiston theory predicted that everything that combusts should decrease in mass, due to the release of phlogiston. However, when metals burn they increase in mass. This is because they combine with oxygen to form a metal oxide, which is heavier than the metal.

Experiment

Zac designed an experiment to measure the mass of 1 g of magnesium (Mg) as it burnt in oxygen gas (O_2) to produce magnesium oxide (MgO). His results are summarised in Table 6.2.1.

Table 6.2.1 Mass of magnesium as measured over 16 seconds as it is burnt

Time (s)	0	1	2	3	4	5	6	7	8
Mass (g)	1.00	1.04	1.08	1.12	1.16	1.20	1.24	1.29	1.34

Time (s)	9	10	11	12	13	14	15	16
Mass (g)	1.39	1.44	1.49	1.55	1.61	1.67	1.67	1.67

1. Plot the data points on the axes below to construct a graph and draw a line of best fit.

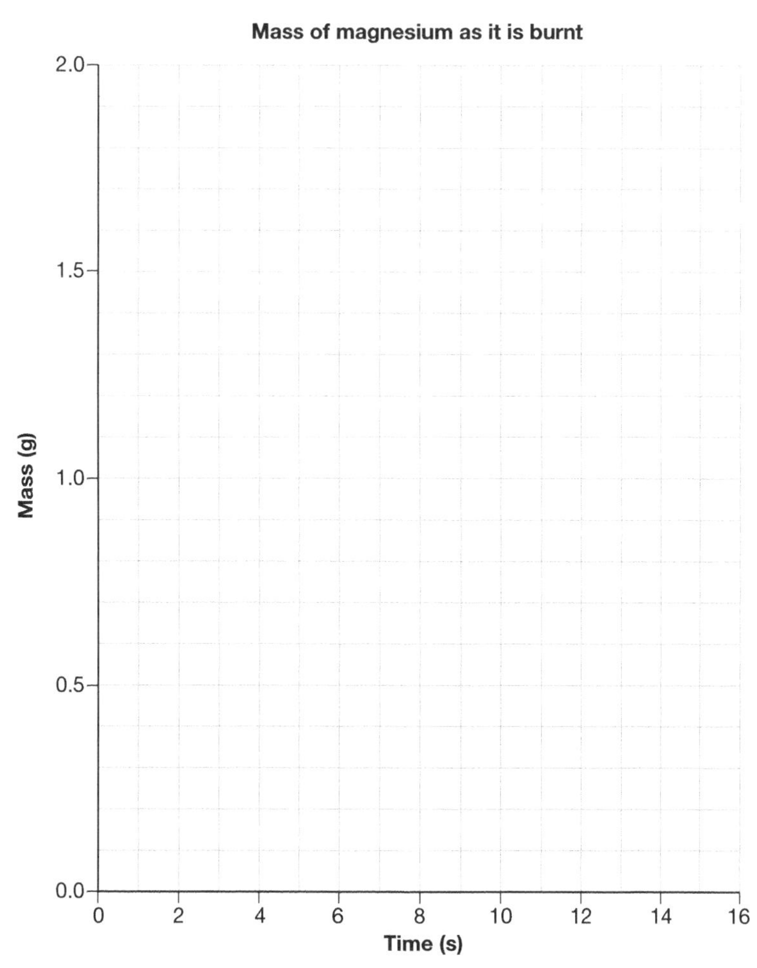

6.2 Combustion of magnesium

(2) Construct particle model diagrams showing the atoms in the reactants and products before and after the reaction in Zac's experiment.

Before	After
O_2 and Mg	MgO

(3) Construct a word equation and a formula equation for the chemical reaction.

(a) word equation: ____________________

(b) formula equation: ____________________

(4) Describe how the mass changed as the magnesium burnt in oxygen.

____

____

(5) **(a)** From the graph, state how much time the chemical reaction took to complete.

____

(b) Explain your answer.

____

____

(6) Calculate the mass of oxygen that was used in the reaction. __________ g

(7) The chemical formula for magnesium oxide, MgO, indicates that there are equal numbers of magnesium and oxygen atoms in the compound. Calculate how much heavier a magnesium atom is compared to an oxygen atom using the equation:

$$\frac{\text{Mass of Mg in MgO}}{\text{Mass of O in MgO}} = \underline{\qquad\qquad} = \underline{\qquad\qquad}$$

(8) Given that the atomic mass of an oxygen atom is 16, calculate the atomic mass of a magnesium atom. Do the calculation in the space below.

RATE MY UNDERSTANDING
Shade the face that shows your rating

6.3 Classifying chemical reactions

Science understanding

FOUNDATION | **STANDARD** | ADVANCED

Scientists classify chemical reactions into different types. The classification might depend on the type of reactants that are involved, the type of products that are produced or how the reactants interact to form the products. This helps scientists to understand more about chemical reactions because by understanding one chemical reaction, they can also understand other chemical reactions of the same type.

Some common types of chemical reactions are:

- *decomposition*—when one reactant breaks apart to form two or more products
- *precipitation*—when two aqueous solutions mix to produce a solid
- *neutralisation*—when an acid reacts with a base to produce a salt and water
- *combustion*—a highly exothermic reaction in which a substance combines with oxygen to produce heat and light.

1. Classify each reaction listed below by placing a cross in the relevant column. Remember, a chemical reaction may fit more than one classification.

Chemical reaction	Decomposition	Precipitation	Neutralisation	Combustion
(a) $4HNO_3 \rightarrow 2H_2O + 4NO_2 + O_2$				
(b) $2AgNO_3 + Na_2S \rightarrow Ag_2S + 2NaNO_3$				
(c) $2H_2 + O_2 \rightarrow 2H_2O$				
(d) $HNO_3 + KOH \rightarrow KNO_3 + H_2O$				
(e) $2Na + O_2 \rightarrow Na_2O$				
(f) $2SO_3 \rightarrow 2SO_2 + O_2$				
(g) $2KI + Pb(NO_3)_2 \rightarrow 2KNO_3 + PbI$				
(h) $H_2SO_4 + Mg(OH)_2 \rightarrow MgSO_4 + 2H_2O$				
(i) $CH_4 + 2O_2 \rightarrow CO_2 + 2H_2O$				
(j) $BaCl_2 + Na_2SO_4 \rightarrow BaSO_4 + 2NaCl$				
(k) $H_2O_2 \rightarrow H_2 + O_2$				
(l) $NaOH + HCl \rightarrow NaCl + H_2O$				

RATE MY UNDERSTANDING
Shade the face that shows your rating

6.4 Classifying chemical reactions in context

Science understanding

FOUNDATION | **STANDARD** | ADVANCED

1. Match the following scenarios with the type of chemical reaction that explains the observations by drawing a line between them.

Scenario	Reaction Type
Trent proudly polished his silver cutlery until it sparkled and then put it away for a special occasion. That occasion came one year later, but when he took out the silverware for his guests he found that it had become dull and was covered in a thin, black-brown layer.	decomposition
Sean decided to bake some bread and so bought some dry yeast. He read that the yeast needed to activate by placing it in some warm water. Knowing that yeast was a living organism, Sean decided to give it an extra boost by feeding it some sugar. When he came back he found that the water containing the yeast had developed ugly, brown foam on top.	precipitation
Lauren daydreamed as she looked into the flames of her campfire and felt its warmth on her face.	neutralisation
Vern lived in an extremely industrialised town. The air had become so polluted that the rain was as acidic as lemon juice and dissolved the limestone buildings and statues.	acid–metal
Meixang didn't bother to read the instructions on her bottle of pure hydrogen peroxide. The instructions said the hydrogen peroxide should be refrigerated. Instead, Meixang left it in the sun on her lab bench. When she came back, she found that it had built up a lot of gas inside. Curious to find out what the gas was, she tested it and found that it was a mixture of hydrogen and oxygen gases.	acid–carbonate
Karen had a great meal but ate too quickly. She soon began to feel a burning sensation in her chest. To soothe the burning she drank some heartburn medication.	combustion
At the end of the water purification process, the water looked mostly clean but it still contained dissolved calcium and magnesium compounds that needed to be removed. To do this, Tom added solutions of lime (calcium hydroxide) and soda ash (sodium carbonate).	corrosion
Joan is an artist whose favourite technique is etching. In this technique, she takes a zinc plate covered in wax and scratches the wax away to reveal the zinc underneath. She then places the plate in an acid bath. The acid eats away the exposed zinc but does not attack the parts covered in wax. When the wax is removed, Joan's artwork has been 'etched' into the zinc plate. She can then cover the plate with ink and use it in a printing press to make hundreds of copies of her artwork.	respiration

RATE MY UNDERSTANDING
Shade the face that shows your rating   

6.5 Reaction rate and surface area of reactants

Science inquiry skills

FOUNDATION | **STANDARD** | ADVANCED

Processing & Analysing **Communicating**

Katrina set up an experiment to compare how crushing a soluble aspirin tablet affects the rate at which it dissolves. She placed a whole aspirin tablet (control) and a crushed aspirin tablet in separate bottles of water and measured the volume of carbon dioxide that is produced over 3 minutes. Her results are summarised in Table 6.5.1.

Table 6.5.1 Results of experiment comparing rate tablet dissolves in water

Time (min)	0.0	0.2	0.4	0.6	0.8	1.0	1.2	1.4
Volume of CO_2 Control (L)	0.00	0.08	0.16	0.26	0.36	0.46	0.58	0.70
Volume of CO_2 Crushed (L)	0.00	0.70	1.90	1.90	1.90	1.90	1.90	1.90

Time (min)	1.6	1.8	2.0	2.2	2.4	2.6	2.8	3.0
Volume of CO_2 Control (L)	0.84	0.98	1.14	1.31	1.49	1.69	1.90	1.90
Volume of CO_2 Crushed (L)	1.90	1.90	1.90	1.90	1.90	1.90	1.90	1.90

1. Use different symbols to construct two graphs from the data in Katrina's experiment on the axes below. Remember to provide a key to show which symbol matches which dataset.

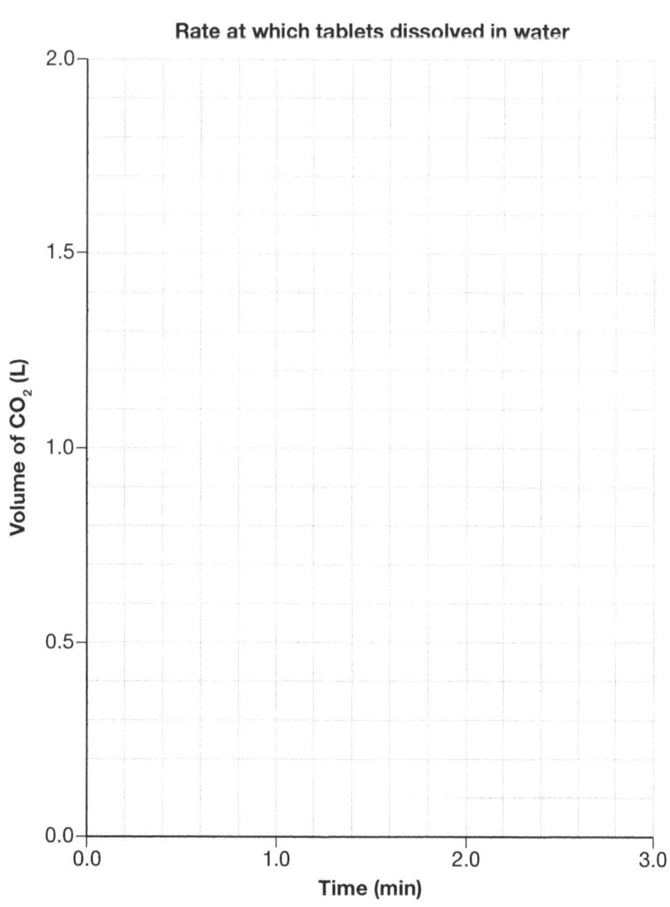

6.5 Reaction rate and surface area of reactants

(2) Compare how the whole and crushed aspirin tablets reacted when placed in water.

(3) State how much carbon dioxide gas was produced by each tablet.

(4) From the graph, calculate the time it took for the:

(a) whole tablet to react completely

(b) crushed tablet to react completely.

(5) Calculate how much faster the crushed tablet reacted than the whole tablet, using the space below. Use the space below for your calculations.

6 A crushed tablet would help to get the aspirin into the bloodstream quickly and relieve pain faster. Propose a reason why this soluble aspirin tablet is not taken as a powder or in capsules.

7 Katrina learns that carbon dioxide is produced when sodium bicarbonate ($NaHCO_3$) and citric acid ($C_6H_8O_7$) in the aspirin tablet react to form sodium citrate ($NaC_6H_7O_7$), carbon dioxide (CO_2) and water (H_2O). Write the formula equation for this reaction.

RATE MY UNDERSTANDING
Shade the face that shows your rating

6.6 The rate of reaction

Science understanding

FOUNDATION | **STANDARD** | ADVANCED

Some chemical reactions are very fast while others proceed very slowly. The speed at which a chemical reaction takes place is known as the rate of reaction.

1. Identify whether the rate of the chemical reaction is usually increased or decreased in each situation below and describe what is done to control the rate.

(a) baking a cake	**(b)** milk going sour
rate increased ☐ rate decreased ☐ by: ______________________ ______________________ ______________________	rate increased ☐ rate decreased ☐ by: ______________________ ______________________ ______________________
(c) food going stale	**(d)** capsules to relieve a headache
rate increased ☐ rate decreased ☐ by: ______________________ ______________________ ______________________	rate increased ☐ rate decreased ☐ by: ______________________ ______________________ ______________________
(e) fruit ripening on the way to market	**(f)** respiration during a race
rate increased ☐ rate decreased ☐ by: ______________________ ______________________ ______________________	rate increased ☐ rate decreased ☐ by: ______________________ ______________________ ______________________
(g) starting a camp fire	**(h)** converting carbon monoxide from a car exhaust into carbon dioxide
rate increased ☐ rate decreased ☐ by: ______________________ ______________________ ______________________	rate increased ☐ rate decreased ☐ by: ______________________ ______________________ ______________________

RATE MY UNDERSTANDING
Shade the face that shows your rating

6.7 Rate of chemical reaction—particle diagrams

Science understanding

FOUNDATION | **STANDARD** | ADVANCED

Construct diagrams to show how the rate of chemical reactions can be controlled by the temperature, concentration of reactants, surface area of reactants, agitation and catalysts.

Include labels in your diagrams and write a short description to explain what is happening to the particles and how the reaction rate is affected in each situation shown.

1. Temperature

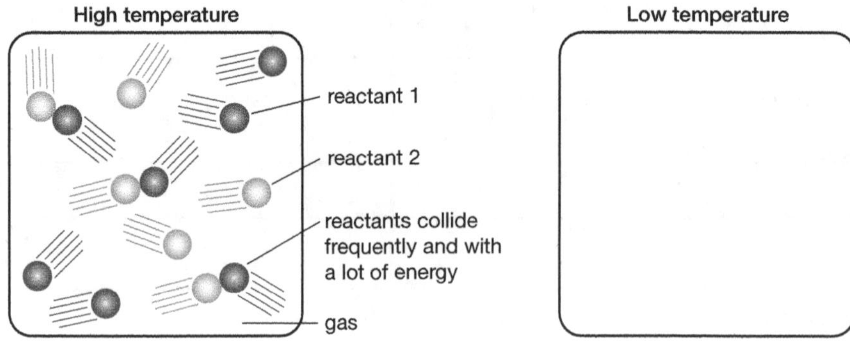

2. Concentration of reactants

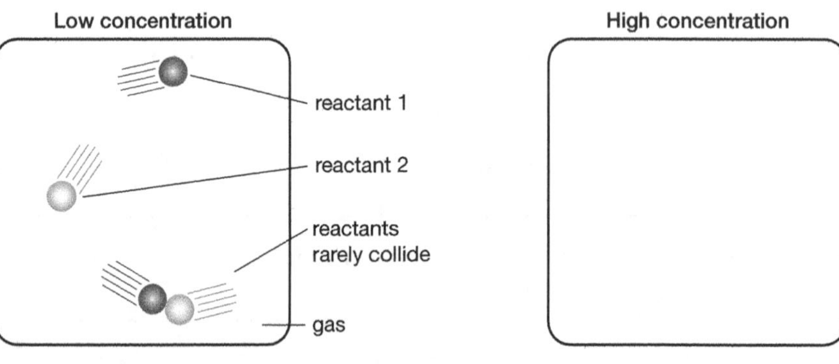

3 Surface area of the reactants

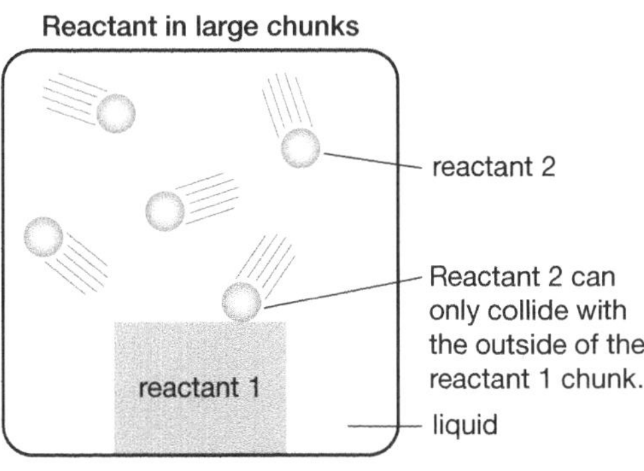

4 Agitation

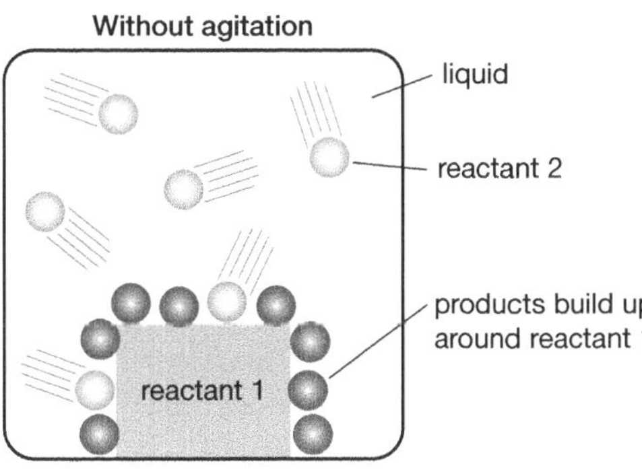

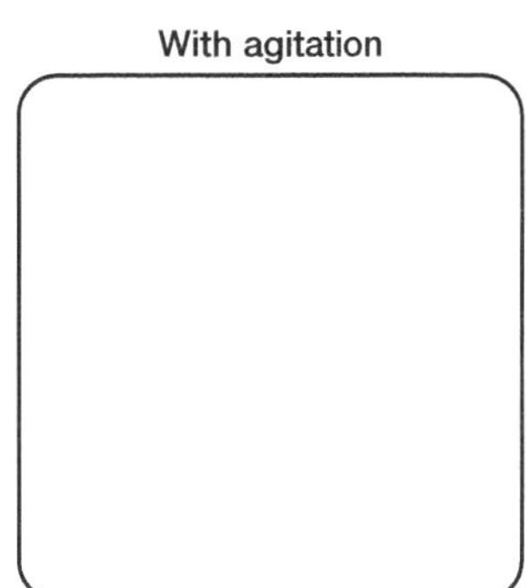

5 Catalysts

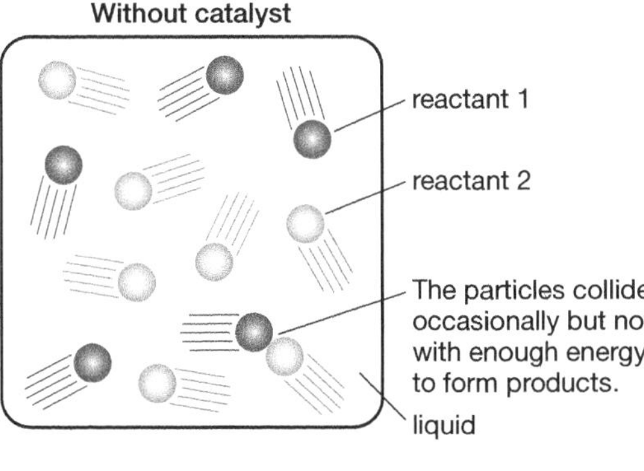

RATE MY UNDERSTANDING
Shade the face that shows your rating

6.8 Balancing chemical equations

Science understanding

FOUNDATION | **STANDARD** | ADVANCED

Atoms are not created or destroyed in a chemical reaction. Therefore, there must be the same number of atoms in the reactants as in the products of a chemical reaction. This law of science is known as the conservation of mass and is shown by the use of balanced chemical equations.

1 State the number of each type of atom in the reactants and products of the following equations. Identify if the equations are balanced.

(a) $C_3H_8 + 5O_2 \rightarrow 3CO_2 + 4H_2O$

propane, oxygen, carbon dioxide, water

Reactants: carbon (C) = ____________

hydrogen (H) = ____________

oxygen (O) = ____________

Products: carbon (C) = ____________

hydrogen (H) = ____________

oxygen (O) = ____________

Balanced/unbalanced: ____________

(b) $H_2SO_4 + 2NaOH \rightarrow Na_2SO_4 + H_2O$

sulfuric acid, sodium hydroxide, sodium sulfate, water

Reactants: hydrogen (H) = ____________

sulfur (S) = ____________

oxygen (O) = ____________

sodium (Na) = ____________

Products: hydrogen (H) = ____________

sulfur (S) = ____________

oxygen (O) = ____________

sodium (Na) = ____________

Balanced/unbalanced: ____________

2 State whether the following chemical equations are balanced or unbalanced.

(a) SO_3 + H_2O → H_2SO_4

sulfur trioxide, water, sulfuric acid

Balanced/unbalanced: ______________

(b) HCl + Na → NaCl + H_2

hydrochloric acid, sodium, sodium chloride, hydrogen

Balanced/unbalanced: ______________

(c) C_4H_{10} + $6O_2$ → $4CO_2$ + $5H_2O$

butane, oxygen, carbon dioxide, water

Balanced/unbalanced: ______________

3 Produce balanced chemical equations by filling in the blanks.

(a) $3H_2$ + N_2 → ______ NH_3

hydrogen, nitrogen, ammonia

(b) ______ Mg + P_4 → $2Mg_3P_2$

magnesium, phosphorus, magnesium phosphide

(c) ______ HCl + $CuSO_4$ → $H2SO_4$ + $CuCl_2$

hydrochloric acid, copper(II) sulfate, sulfuric acid, copper(II) chloride

(d) 2Al + ______ CuO → Al_2O_3 + ______ Cu

aluminium, copper(II) oxide, aluminium oxide, copper

(e) 3C + ______ HNO_3 → ______ CO_2 + $2H_2O$ + ______ NO

carbon, nitric acid, carbon dioxide, water, nitric oxide

6.9 Literacy review

Science understanding

FOUNDATION	STANDARD	ADVANCED

1 Recall your knowledge of chemical reactions by using the words from the box to complete the paragraphs below.

agitation area base carbon dioxide catalysts catalytic combustion concentration corrosion decomposition endothermic enzymes exothermic explosion heat light metals neutralisation rate respiration rusting salt solid soluble temperature

(a) Scientists classify chemical reactions into different types of reactions. One method of classifying chemical reactions is based on whether they release or absorb energy in the form of ________________ and ________________. Chemical reactions that release energy are known as ________________ reactions. Chemical reactions that absorb energy are known as ________________ reactions. Oxygen gas is often involved in exothermic reactions such as ________________, ________________ and ________________.

(b) Another type of reaction is ________________ where one reactant breaks apart to form several products. Precipitation reactions are reactions where two ________________ compounds in solution mix to form a ________________ precipitate.

(c) Acids are common reactants and so many types of chemical reactions involve the use of acids. ________________ reactions involve the reaction of an acid with a ________________ to produce a ________________ and water. When acids react with ________________ they produce a salt and hydrogen gas. And when acids react with carbonates they produce a salt, water and ________________.

(d) The ________________ of a chemical reaction is how fast the reaction takes place. An example of a fast chemical reaction is an ________________. An example of a slow chemical reaction is ________________.

(e) The rate of chemical reactions can be controlled by:

- controlling the ________________
- controlling the ________________ of the reactants
- controlling the surface ________________ of the reactants
- stirring or mixing, which is also known as ________________
- adding chemical helpers known as ________________.

(f) Nature produces its own chemical helpers called ________________ to increase the rate of chemical reactions in your body.

RATE MY UNDERSTANDING
Shade the face that shows your rating

6.10 Thinking about my learning

1 **(a)** For each of the topics in the table below, tick the box that you feel most represents your understanding.

Topic	I need some help with this	I mostly understand this	I am confident that I understand this
I can write word equations.			
I can write formula equations.			
I can balance chemical equations.			
I can identify states (solid, liquid, gas and aqueous solution) and write them as formula equations.			
I can identify decomposition reactions.			
I can identify combination reactions.			
I can identify precipitation reactions.			
I can identify oxidation reactions.			
I can identify reduction reactions.			
I can identify metal displacement reactions.			
I can predict whether a metal displacement reaction will occur.			
I can name ionic compounds.			
I can predict precipitation reactions.			
I can describe froth flotation.			
I can explain carbon reduction.			
I can explain electrolysis.			
I can describe how temperature impacts the rate of reaction.			
I can describe how concentration impacts the rate of reaction.			
I can describe how agitation impacts the rate of reaction.			
I can describe how surface area impacts the rate of reaction.			
I can describe how catalysts impact the rate of reaction.			
I can describe what an enzyme is.			

(b) From the table above, highlight in red the three topics that you feel the least confident about.

(c) From the table above, circle the three topics that you feel the most confident about.

2 From this chapter, state one thing that you found the most interesting or fun to learn.

Global systems

7.1 Knowledge preview

Science understanding

FOUNDATION | STANDARD | ADVANCED

The Earth is a series of interconnected systems. You may be familiar with some elements of these spheres through discussion in the media, your school studies or through internet exploration. Other elements will be quite new to you.

1 Complete the table below which lists some of the global systems with which you may be familiar. Place a tick in the column which applies to your level of knowledge.

Element of Earth's system	I am unfamiliar with this	I have heard of this but do not know any details	I have heard quite a bit about this
water cycle			
carbon cycle			
causes of cyclones			
the jet stream			
El Niño and La Niña			
ozone hole			
greenhouse effect			
global warming			
climate change			
global conveyor belt			
biodiversity			
sustainable living			

2 Choose three of the elements with which you are familiar. Write two sentences about each element that show your understanding of what they are.

3 The hole in the ozone layer, the greenhouse effect and global warming have been common topics for media discussion over the last 20 years. Explain how you think these are related to each other.

4 Many elements and compounds are naturally cycled through the environment. Why do you think these cycles are important? What do you think would happen if materials were not recycled?

5 Australians have a relatively large carbon footprint.

(a) Explain what the term 'carbon footprint' means.

(b) What are some of the possible reasons why the carbon footprint of Australians is large in comparison to the rest of the world?

7.2 Natural carbon recycling

Science understanding

FOUNDATION | **STANDARD** | ADVANCED

Carbon is the basis of all life on Earth. There is a fixed amount of carbon on Earth and this carbon is continuously recycled and reused. It takes carbon about 100 million years to complete one cycle through the lithosphere (rock, soil, land), hydrosphere (oceans, lakes, rivers), the atmosphere and the biosphere.

bushfire (*n*) a forest fire that is out of control

controlled burn (*n*) a supervised fire

Carbon stores

The Earth's four major carbon stores (or sinks) are—the atmosphere, oceans, terrestrial ecosystems (on land) and Earth's crust.

HINT

1 **Gigatonne** = 10^9 tonne = 10^{12} kg

Table 7.2.1 Earth's principal carbon stores in gigatonnes (Gt)

Stores of carbon	Gigatonnes (Gt)	Stores of carbon	Gigatonnes (Gt)
marine sediments and sedimentary rocks	66 000 000–100 000 000	coal	3000
deep oceans	40 000	soils and organic matter	1600
surface oceans	1000	oil and gas	300
vegetation	540–610	the atmosphere	750

Plants and the carbon cycle

Plants absorb carbon dioxide from the atmosphere through photosynthesis. Some carbon is returned to the atmosphere almost immediately through respiration. The remainder is used to build leaves, wood and roots, and may be stored for hundreds of years either as the tree itself or in wood products made from it, such as the frames of houses or furniture.

Plant litter (fallen leaves and twigs) and roots become part of the soil when decomposer organisms break them down. As the litter is broken down, carbon is released into the atmosphere.

Bushfires, controlled burns or using firewood to heat homes rapidly releases stored carbon into the atmosphere. Most of the carbon will be in the form of carbon dioxide. There may also be carbon monoxide and methane, depending on the conditions of burning. Following burning, carbon may also be added to the soil in the form of charcoal, which remains stable for many years.

1 Name the Earth's four major stores of carbon.

______ ______ ______ ______

2 Use the data about carbon stores in Table 7.2.1 and rewrite the carbon stores, ranking them from the largest to the smallest storage.

Largest ______ ______

______ ______

______ ______

______ Smallest ______

3 Which do you think is the most stable store? Explain.

______

______

4 Which store do you think changes most rapidly? Explain.

______

______

RATE MY UNDERSTANDING
Shade the face that shows your rating

7.3 The carbon cycle

Science understanding

FOUNDATION | **STANDARD** | ADVANCED

Human activity changes the carbon cycle

Understanding the way carbon is cycled in nature is helpful when looking at ways human activities have affected the carbon cycle. A significant increase in the amount of gases containing carbon compounds in the atmosphere is thought to be caused by a range of human activities including:

- burning of fossil fuels (coal, oil and natural gas) as a source of energy for industry, transport and households—releasing CO_2 and CH_4
- deforestation (removal of trees)—releasing CO_2 and reducing CO_2 uptake through photosynthesis
- changing land use (for example, for agriculture or urban development)—releasing CO_2
- production of cement and lime—releasing CO_2 from very long-term stores
- waste burning and waste decomposition in landfills—releasing CO_2 and CH_4
- livestock and rice cultivation—releasing CO_2 and CH_4.

cement (*n*) a powder made from clay and limestone used to make concrete

cultivation (*n*) the preparation of soil for farming

degraded (*adj*) poor quality

emission (*n*) something that is sent out or released

landfill (*n*) large holes filled with rubbish

lime (*n*) a calcium compound, used to make cement

Of these activities, burning fossil fuels is responsible for about three-quarters of the increased carbon emissions.

About half the additional carbon emissions produced by human activity remains in the atmosphere. The rest is taken up by carbon stores in the oceans and terrestrial ecosystems in about equal proportions.

Animals and the carbon cycle

When animals consume plants, they convert some of the plant carbon into the tissues of their bodies. Some carbon is returned to the atmosphere through the animals' respiration and some is added to the soil as wastes which are used by decomposer organisms as a food source.

Agriculture and the carbon cycle

The concentration of carbon dioxide and methane in the atmosphere is influenced by agricultural activities. For example:

- Cultivating soils can increase the rate of breakdown of soil and organic matter, resulting in higher emissions of carbon dioxide to the atmosphere.
- Ruminant or cud-chewing animals such as cattle also return some carbon to the atmosphere as methane.
- Land that is in good condition generally stores more carbon in the soil than degraded land with less vegetation cover.

(1) In many areas, forests are burnt to clear land for agriculture. Explain how this method of deforestation would alter the carbon cycle.

7.3 The carbon cycle

(2) Name the human activity that is responsible for the greatest change to the amount of carbon available for cycling.

____

(3) Explain how farming cattle to provide meat and milk may change the amount of carbon in the atmosphere.

____

____

____

____

____

(4) Analyse the information provided about how human activity changes the carbon cycle. On the diagram of the carbon cycle below:

(a) colour in red the arrows that represent movements of carbon that would be increased by human activity

(b) colour in blue the arrows that represent movements of carbon that would be reduced by human activity

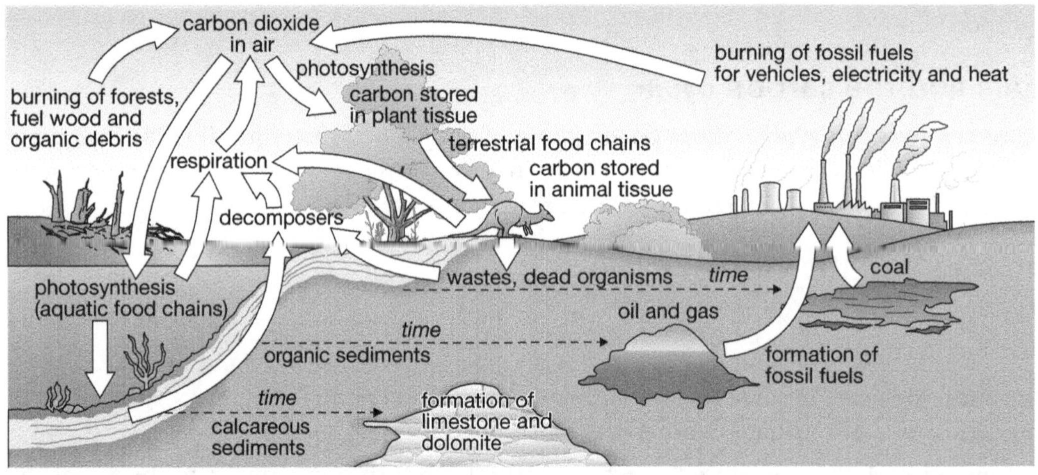

(c) number each coloured arrow and explain the human activity involved and how the activity would bring about change.

____

____

____

____

____

____

RATE MY UNDERSTANDING
Shade the face that shows your rating

7.4 Southern Oscillation

Science understanding

FOUNDATION | STANDARD | **ADVANCED**

Australia is a land of weather extremes. The Southern Oscillation is one cause and it gives rise to the El Niño and La Niña effects.

The Southern Oscillation is a sequence of changes to the way the atmosphere and water circulate across the Pacific Ocean and Indonesian islands. In most years, a cold current flows northwards along the coast of South America, then westwards along the equator, where it is warmed by the Sun. This 'normal' situation is shown in Figure 7.4.1. The result is a difference of 3°C to 8°C between the cooler eastern Pacific and the warmer western Pacific.

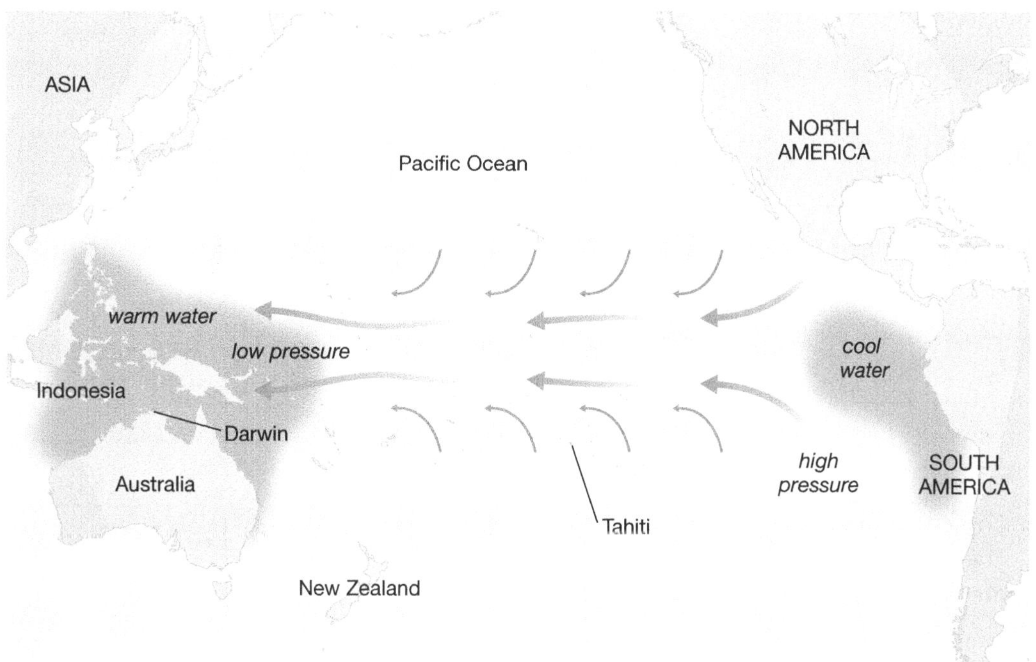

Figure 7.4.1 In normal conditions where there is no El Niño or La Niña, the trade winds blow strongly from the eastern Pacific Ocean to the west. This brings average rainfall to northern Australia.

El Niño is one extreme of the Southern Oscillation and is probably one of the most important influences on the climate in Australia, particularly in Queensland and much of New South Wales. In El Niño years, there may be little or no difference in temperature between the western and eastern Pacific. With little temperature difference, there is also little pressure difference. The trade winds that normally blow strongly from South America weaken and the winds carrying a lot of moisture do not reach Australia. Cool air descends over Australia bringing little rainfall. During an El Niño event, large areas of Australia may experience drought.

The opposite of El Niño is La Niña. During a La Niña, the central and eastern Pacific Ocean becomes much cooler than normal. The trade winds blow more strongly than usual and Australia experiences more cloud and wetter-than-normal conditions, especially in the north. When La Niña takes over, Australia often experiences floods.

7.4 Southern Oscillation

1. Figure 7.4.2 illustrates the conditions during an El Niño event. Describe what is happening at each position labelled A to E.

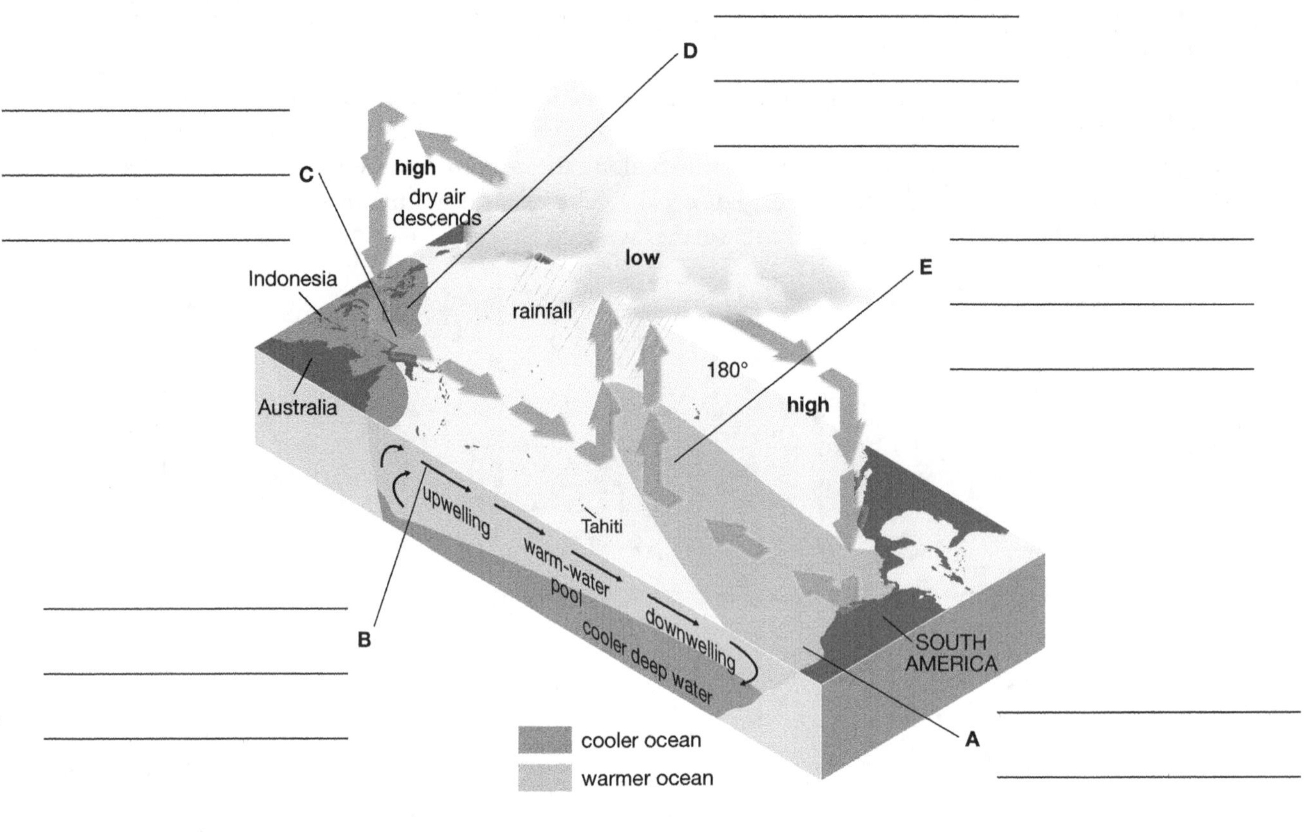

Figure 7.4.2 Climatic conditions in an El Niño year

2. Compare the events illustrated in Figure 7.4.2 with a 'normal' weather pattern.

3. Figure 7.4.3 illustrates the conditions over the Pacific Ocean when there is a La Niña event. Describe what is happening at each position labelled M to Q.

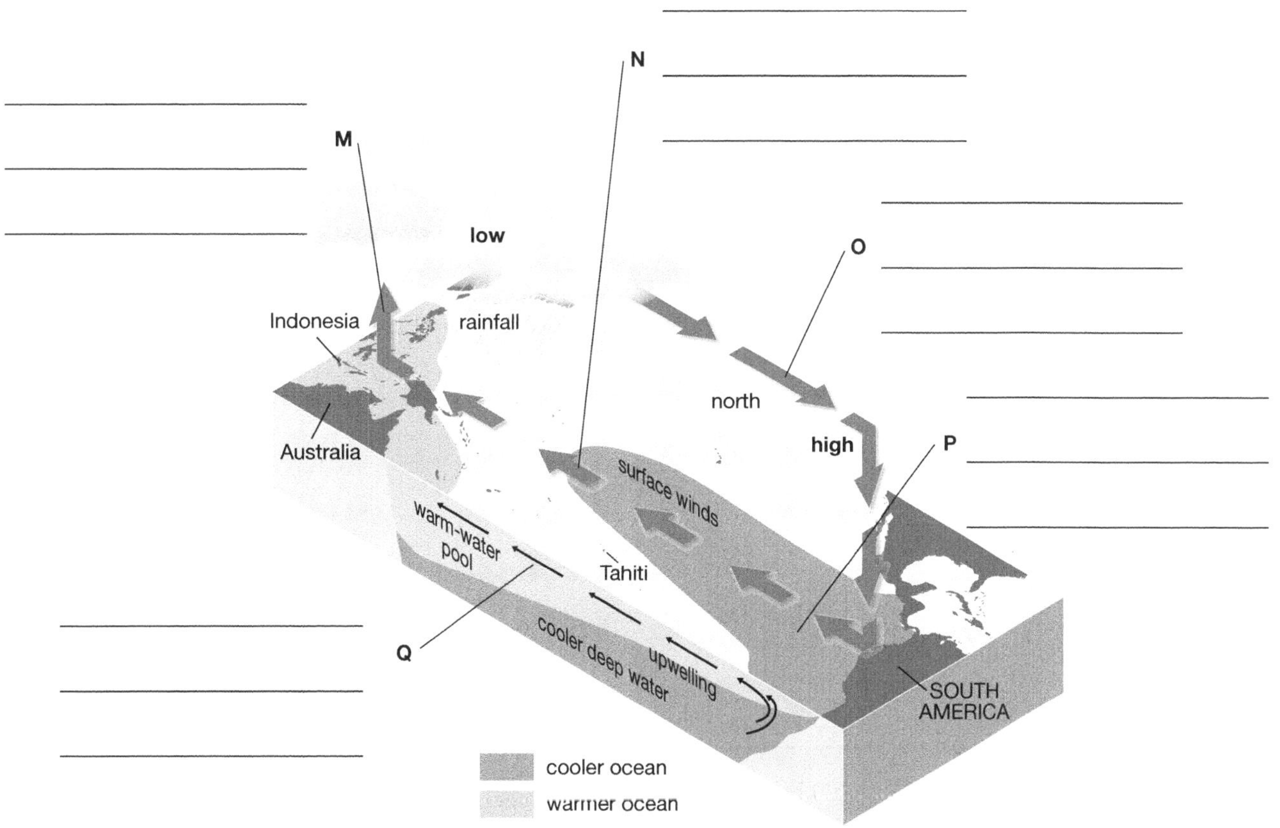

Figure 7.4.3 Climatic conditions in a La Niña year

4. Compare the events illustrated in Figure 7.4.3 with a 'normal' weather pattern.

7.5 Climate graph

Science inquiry skills

FOUNDATION	STANDARD	ADVANCED

Processing & Analysing | Communicating

The Southern Oscillation Index (SOI) is a measure of the atmospheric and ocean conditions across the Pacific Ocean (Figure 7.5.1). It is calculated using the difference in air pressure between Tahiti and Darwin. Under 'normal' conditions the SOI is close to zero. During El Niño events, the SOI is strongly negative. During La Niña, the SOI is strongly positive.

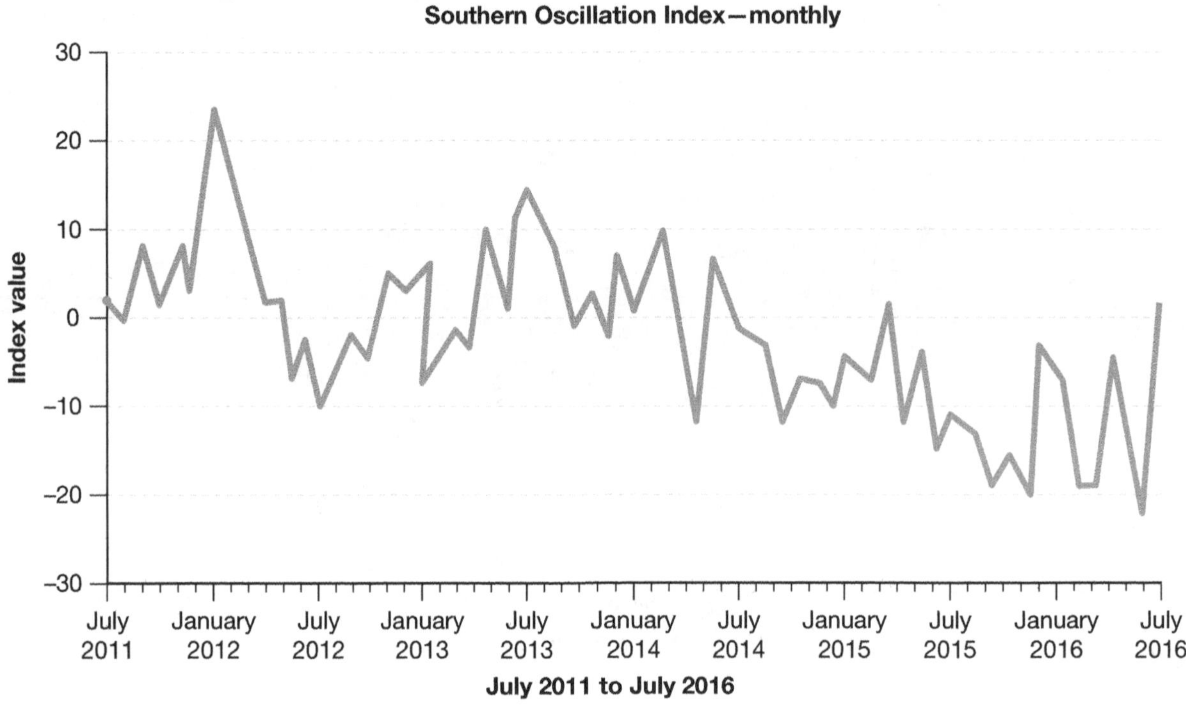

Figure 7.5.1 Southern Oscillation Index from July 2011 to July 2016

1 Describe the weather conditions in Australia when the SOI indicates an El Niño event.

(2) Use a red highlighter to shade the periods of El Niño experienced in Australia between July 2011 and July 2016.

(3) Comment on how you made your decision in question 2.

4 Describe the weather conditions experienced in Australia during a La Niña event.

____

____

____

5 Use a blue highlighter to shade the periods of La Niña experienced in Australia between July 2011 and July 2016.

6 Explain how you made your decision in question 5.

____

____

7 Using evidence of the trend in the SOI from July 2015 to July 2016, suggest how the SOI will change in the next three years. Provide evidence for your answer.

____

____

____

____

____

____

____

RATE MY UNDERSTANDING
Shade the face that shows your rating

7.6 Carbon dioxide emissions

Science inquiry skills

FOUNDATION | **STANDARD** | ADVANCED

Processing & Analysing | Communicating

1. Construct four line graphs on the grid below using the data in Table 7.6.1. Remember to provide a heading and a key to show which line graph matches which country. Different colours may be used for the graph of each country.

Table 7.6.1 CO_2 emissions by country

Year	CO_2 emissions (tonnes per head of population)			
	Australia	USA	China	Japan
1960	8.60	16.20	1.20	2.50
1965	10.60	17.70	0.70	3.90
1970	11.40	20.60	0.90	7.10
1975	11.90	19.70	1.20	7.60
1980	14.80	20.30	1.50	7.90
1985	15.20	18.50	1.90	7.50
1990	17.20	19.20	2.10	8.70
1995	17.10	19.50	2.70	9.00
2000	17.10	20.00	2.60	9.50
2005	18.10	19.50	4.30	9.60
2007	18.10	19.50	4.30	9.60
2011	16.50	17.00	6.70	9.30
2015	16.70	17.00	7.00	9.00

CO_2 emissions (tonnes per head of population)

Year

7.6 Carbon dioxide emissions

(2) Explain why 'CO_2 emissions (tonnes per head of population)' is used as a measure of energy consumption. ____

(3) Compare the CO_2 emissions per head of Australia and the United States of America.

(4) Describe the trend in energy use in China. ____

(5) Compare the data for emissions per head from the United States of America with the data for the other three countries. ____

(a) Rank the countries from highest to lowest carbon emissions producer per head, for the year 2015. ____

(b) The table below provides the total populations for the four countries in the year 2015. Complete the table by first filling in the carbon emissions per head for each country for 2015, then calculating the total carbon emissions for the country in 2015. The first has been completed for you.

Total populations 2015: Australia, USA, China and Japan		
Country—population	**Carbon emissions per head in tonnes**	**Total carbon emissions in tonnes**
Australia—23 million	16.7 million	23 million × 16.7 = 384 million tonnes
USA—319 million		
China—1375 million		
Japan—126 million		

(c) Compare your rankings for carbon emissions per head and total carbon emissions. Explain the differences in the two sets of rankings.

7.7 Climate change

Science inquiry skills

FOUNDATION | **STANDARD** | ADVANCED

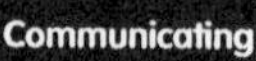

There is discussion in the media, in government and in the scientific community about the effect of human activities on the world climate. Climate is not static. World climate has changed in the past and will continue to change.

Scientists cannot know what the world climate is going to be like in the future. However, they can construct models of what they think will happen.

Computer models of climate change use quantitative data relating to interactions between the atmosphere, oceans, land surface and ice. The models predict the effects of increased amounts of greenhouse gases in the atmosphere. All the models project an upward trend in temperature. However, not all models predict the same amount of change.

greenhouse gas (*n*) gases such as carbon dioxide that can cause the atmosphere to become warmer

quantitative (*adj*) measured using numbers

static (*adj*) not changing, fixed

upward trend (*n*) increase over time

Temperature predictions

Models can take into account different levels of greenhouse gas emissions.

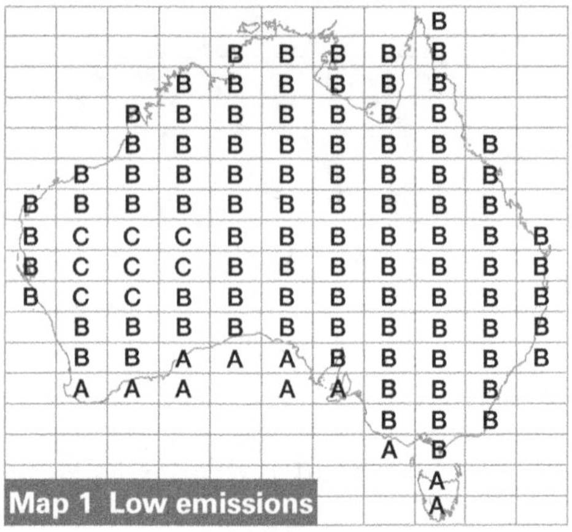

1. Use maps 1 to 3 to demonstrate different temperature predictions for Australia based on low emissions, medium emissions and continued high levels of greenhouse gas emissions.

   Decide which colours to use and colour in the key below. Generally lower temperatures will be shown in blues and greens while higher temperatures will be in yellows and reds.

   Using the key, colour the maps to show the projected change in average temperature from 1990 to 2050. If your key is green for temperatures between 0.6°C and 1°C, then all the letters A on the map will be coloured green.

Key:

		A	B	C	D	E			
	0.3	0.6	1	1.5	2	2.5	3	4	5

Temperature rise (°C)

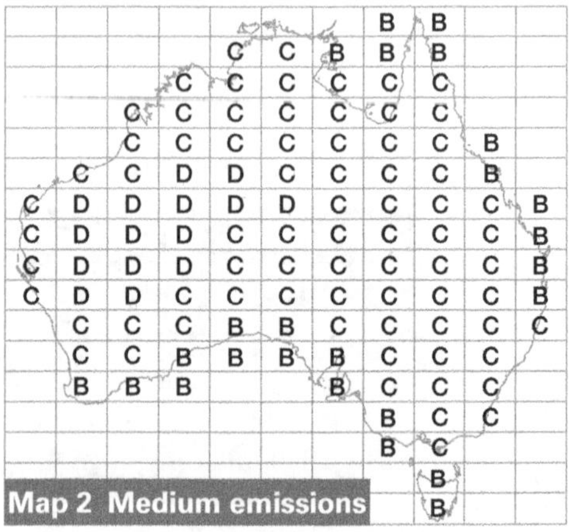

Map 1 Low emissions

Map 2 Medium emissions

Map 3 High emissions

2. Suggest the reasons for using different levels of greenhouse gas emissions in the models for climate change.

____

____

____

____

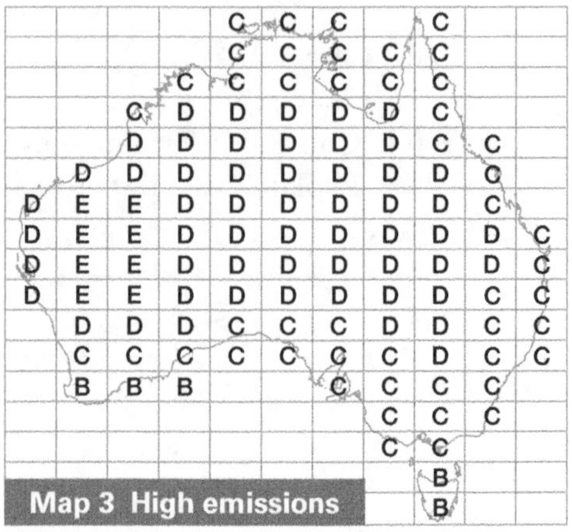

Rainfall predictions

There are many different climate models. Different models can give different results depending on how the data has been used.

7.7 Climate change

Maps 4 to 6 all represent medium levels of emissions of greenhouse gases to predict the percentage change in rainfall from 1990 to 2050. Map 5 represents the mid-point of the model results. Maps 4 and 6 represent the extremes of low and high percentage change in rainfall predicted by different models.

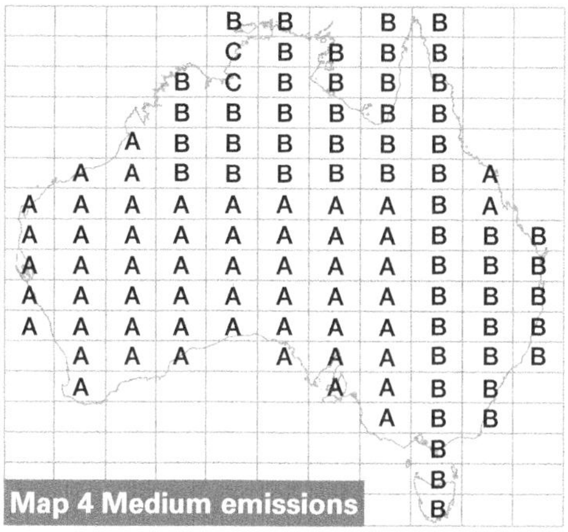

(3) Use maps 4 to 6 to demonstrate how projected changes in average rainfall vary according to different models. Colour the key for each of the percentage changes in rainfall A to J.

Key:	A	B	C	D	E	F	G	H	I	J

−40 −30 −20 −10 0 10 20 30 40

Percentage change in rainfall (%)

Colour the maps to show the projected change in average rainfall from 1990 to 2050.

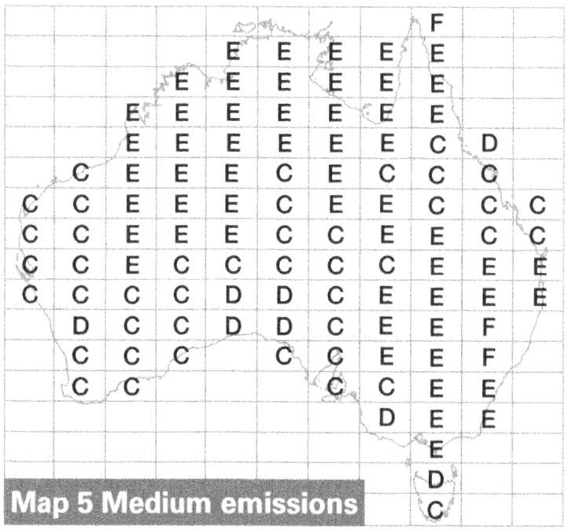

(4) Propose reasons why the different models produce widely different projected changes to rainfall.

____

____

____

____

Map 4 Medium emissions

Map 5 Medium emissions

Map 6 Medium emissions

5 Discuss problems governments could face planning future strategies when presented with such different projections.

____

____

____

____

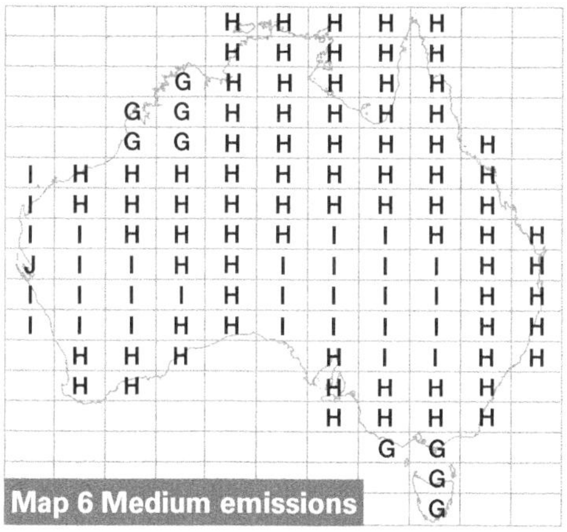

6 Mark where you live on maps 1–6. Identify the projected change in temperature for low, medium and high (maps 1, 2 and 3) levels of greenhouse gas emission.

Low ______ Medium ______ High ______

7 Using map 5, identify the projected percentage change in rainfall for your area.

____

8 Outline how you think the changes in temperature and rainfall would affect life in your area.

____

____

____

7.8 Australian Alps and climate change

Science understanding

FOUNDATION | **STANDARD** | ADVANCED

Australian alpine environments cover less than 0.15% of the continent. Figure 7.8.1 shows the distribution of alpine environments through southern New South Wales, eastern Victoria and central western Tasmania. During the winter, alpine areas are the coldest environments in Australia.

In the highest parts of the Australian Alps around Mt Kosciuszko (2228 metres), the average temperature of the warmest month is less than 10°C and snow is present for about 120 days each year. It is too cold there for trees to grow. At these low temperatures and with snow covering their leaves, plants cannot photosynthesise to convert carbon into woody growth.

alpine (*n*) high mountain

hibernate (*v*) to sleep and slow down the body's processes during winter

metabolism (*n*) the process of burning food for energy

pygmy (*adj*) very small

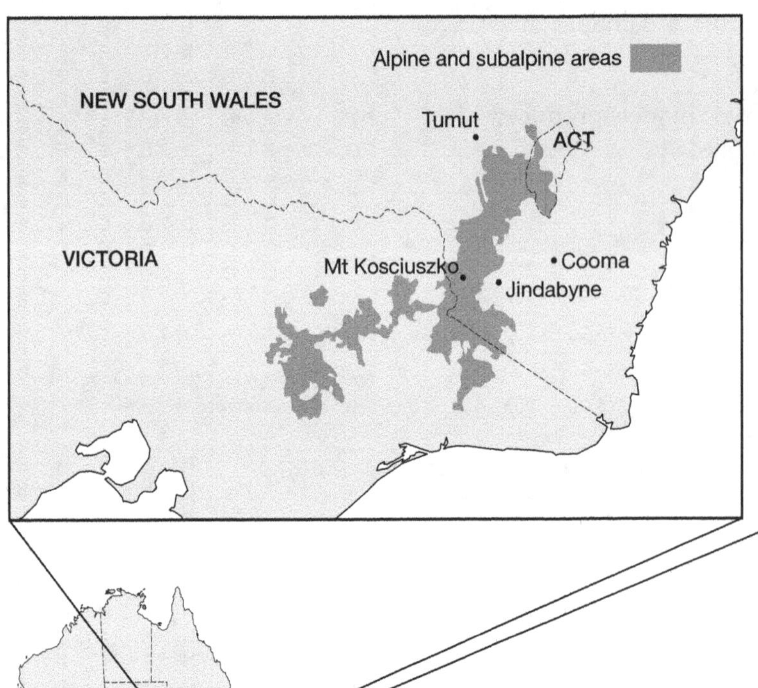

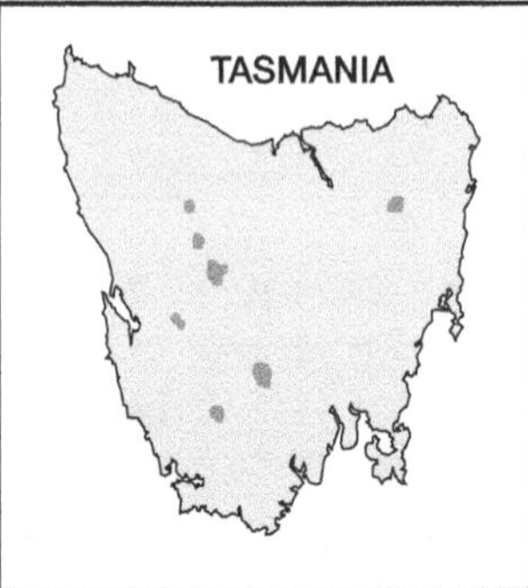

Figure 7.8.1 The alpine areas of Australia

Alpine plants

Alpine vegetation consists of herb fields, grasslands and heath. Most of the plants of the Alps belong to the same genera growing in other areas of Australia. However, the alpine species have evolved special characteristics in response to the harsh environment. The Mt Kosciuszko alpine area is home to 21 species of plants that grow nowhere else in the world.

1 The Australian Alps are areas at the top of high mountains where no trees grow. Explain why trees are not found at these high altitudes.

(2) Suggest what changes could occur in these areas as the world temperature increases.

____

____

____

(3) There are many unique plant species in the Mt Kosciuszko National Park. What would happen to them as the climate changes?

____

____

____

Unique animals

The unique vegetation creates unique animal habitats and some animals can only live in the Australian Alps.

The mountain pygmy-possum (*Burramys parvus*; Figure 7.8.2) is the only marsupial that truly hibernates over the winter. Other marsupials may sleep and be inactive but they don't lower their body temperature to nearly zero and stop their metabolism, the signs of full hibernation. Fossil records show that in the past, when Australia's climate was cooler, the mountain pygmy-possum lived in a more widespread area. It is now restricted to living in the mountains above 1400 metres. The possum shares these high altitudes with just four other mammals.

altitude (*n*) height above sea level
biodiversity (*n*) the variety of plant and animal species
endangered (*adj*) in danger of dying out
fragile (*adj*) easily destroyed
graze (*v*) to eat grass

Figure 7.8.2 The mountain pygmy-possum feeds on stored seeds when it comes out of hibernation and before migrating Bogong moths arrive. The possum then feeds on the moths, building up a fat store that lasts it through hibernation.

Figure 7.8.3 The corroboree frog has black and yellow stripes. It is well camouflaged in the sphagnum moss where it lays its eggs.

An amphibian found only in the New South Wales and Australian Capital Territory Alps is the black-and-yellow-striped corroboree frog (*Pseudophryne corroboree*; Figure 7.8.3). It is only 3 cm long and hibernates under the snow in winter. It lays its eggs in a burrow in sphagnum moss.

4 Identify the effect of climate change on the mountain pygmy-possum in the past.

5 Propose the effect of current and future climate change on the mountain pygmy-possum and black-and-yellow-striped corroboree frog.

Climate change

Recent modelling of climate change predicts that the average temperature of the warmest month in the Australian Alps will be greater than 10°C. This means that trees will be able to grow in the area now occupied by alpine vegetation.

Even where trees do not expand their range, woody shrubs will move in. The leaves of these shrubs will shade the herbs and grasses of the alpine communities and prevent their growth. The result will be a loss in biodiversity.

Changes in the Australian Alps have already been recorded. Over the last 30 years, as temperatures have warmed slightly, kangaroos, hares and wild horses are being found in larger numbers at high altitudes. These animals graze on the fragile alpine vegetation, damaging its habitat and reducing plants numbers. Seven out of eleven species of migratory birds that fly to the Alps annually are arriving earlier. Foxes are moving uphill into the Alps. Bogong moths are arriving later, forcing the mountain pygmy-possums to travel more widely to find food. This increases the opportunities for foxes to prey on the already endangered possums. Foxes also eat Bogong moths. Now that the foxes and possums live in the same area, they are in direct competition for the moths as a food source.

6 Explain how increasing temperatures could cause a reduction in biodiversity worldwide.

RATE MY UNDERSTANDING
Shade the face that shows your rating

7.9 Sea level changes

Science inquiry skills

FOUNDATION | STANDARD | ADVANCED

Processing & Analysing | Evaluating

Figure 7.9.1 illustrates changes in sea level for the last 450 000 years.

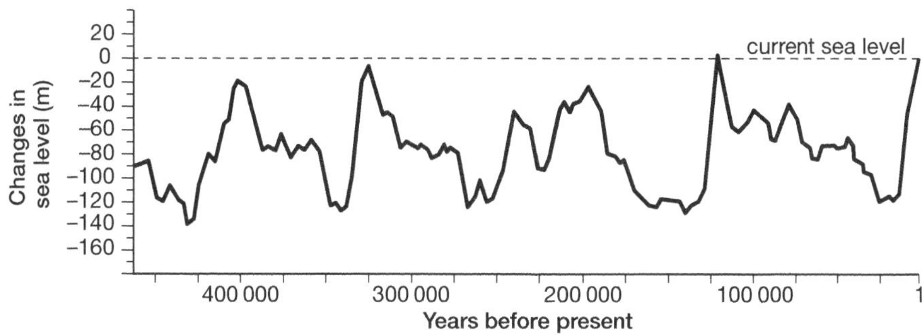

Figure 7.9.1 Historic changes in sea level—late Pleistocene and Holocene. Trend over 450 000 years.

1. Analyse the data, then mark on the graph times when ice ages occurred in the last 400 000 years.

2. Explain why you chose to mark those times.

____

____

3. When did the latest rise in sea level start to occur?

____

4. Figure 7.9.2 has two separate curves, both representing global changes in sea level but using data from different sources.

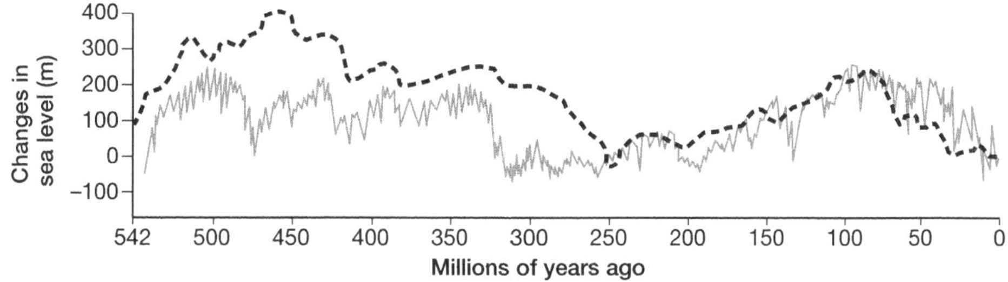

Figure 7.9.2 Global changes in sea level—trend over 540 million years

(a) Compare Figures 7.9.1 and 7.9.2.

____

____

(b) Explain why the two graphs appear so different.

____

____

(5) Suggest reasons for scientists reaching different conclusions about sea levels in the past.

6 What would you say to someone who suggests that the sea levels we are experiencing are among the highest ever?

Figure 7.9.3 is also a graph showing the change in sea level.

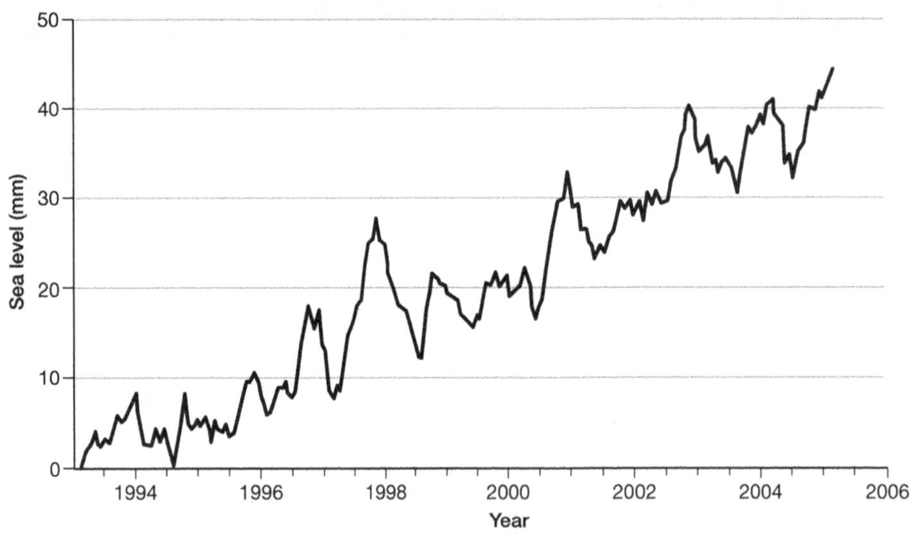

Figure 7.9.3 Changes in sea level from 1993 to 2006

(7) From the shape of the graph, what would you conclude about the changes occurring?

(8) Explain how the scale on the graph in Figure 7.9.3 could make people believe that a fast and large rise in sea level is occurring.

(9) Calculate the rise in sea level in the 10 years before 2006.

(10) Calculate the rise in sea level in the 20 000 years since the last ice age.

RATE MY UNDERSTANDING
Shade the face that shows your rating

7.10 Impact of rising sea level

Science inquiry skills

FOUNDATION | STANDARD | ADVANCED

Questioning & Predicting | Processing & Analysing

This is a map of a busy coastal town. Dairy farming and tourism are the major industries of the town, with large butter and cheese factories in the industrial area. Beef cattle are also important but there is no meat processing in the area. The cattle have to be shipped by road or rail to major centres north of the city.

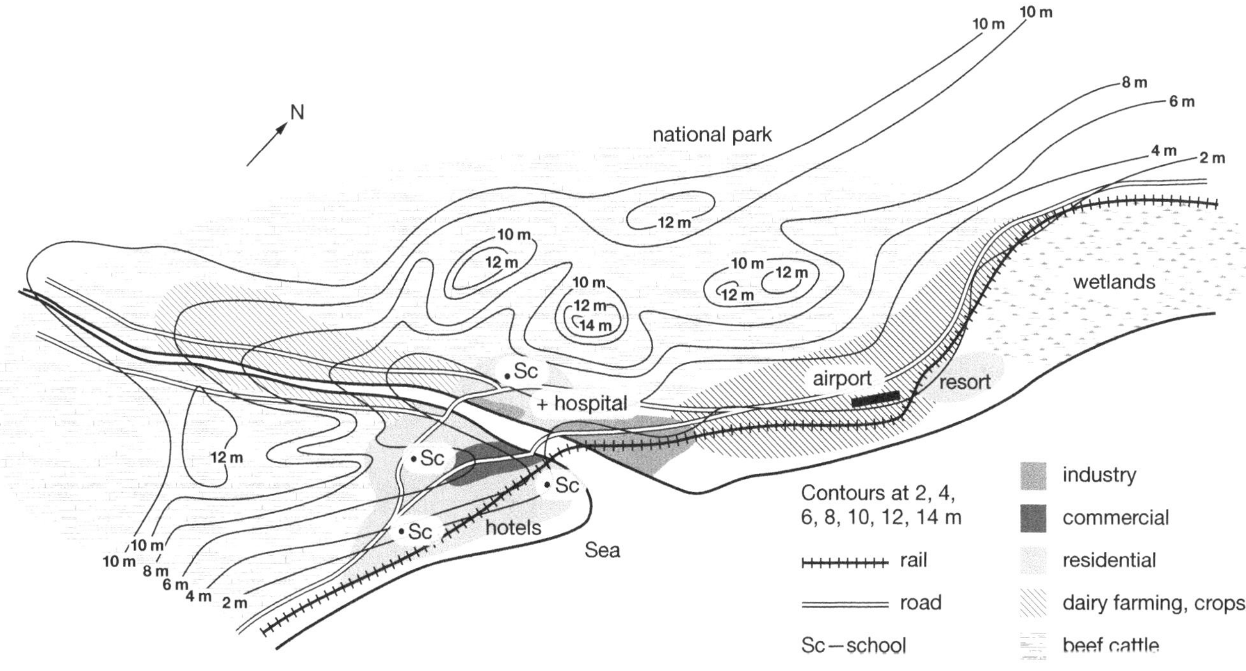

1 Predict the impacts on this city if the sea level rose by the following heights. It may be helpful to colour the areas that will become flooded as sea level rises.

(a) 1 metre ________________________________

(b) 2 metres ________________________________

(c) 6 metres ________________________________

RATE MY UNDERSTANDING
Shade the face that shows your rating

7.11 Literacy review

Science understanding

FOUNDATION | **STANDARD** | ADVANCED

1 Recall your knowledge of global systems by writing the term described by each of the following definitions on the grid. If all the terms are correct, then the highlighted letters should spell a word relevant to this chapter.

(a) The gas that is fixed in the root nodules of leguminous plants

(b) Units used to measure the thickness of the ozone layer

(c) The land masses on Earth

(d) ________________ gases trap heat close to Earth's surface

(e) The word representing the 'S' in SOI—an indication of the difference in air pressure between Tahiti and Darwin

(f) Circular patterns shown by the ocean currents in the major ocean basins

(g) Fuels containing the carbon of plants and animals that died and were preserved millions of years ago

(h) Areas on Earth where the temperature in layers of soil or rock beneath the surface never rise above freezing point

(i) The process by which carbon is recycled through the soil, through living things and the atmosphere

(j) All living things on Earth

(k) An extreme of the Southern Oscillation that causes significant rainfall in parts of Australia

(l) The long-term averages of weather conditions

(m) Term used to describe ecosystems that are diverse and provide for the needs of the organisms that live there

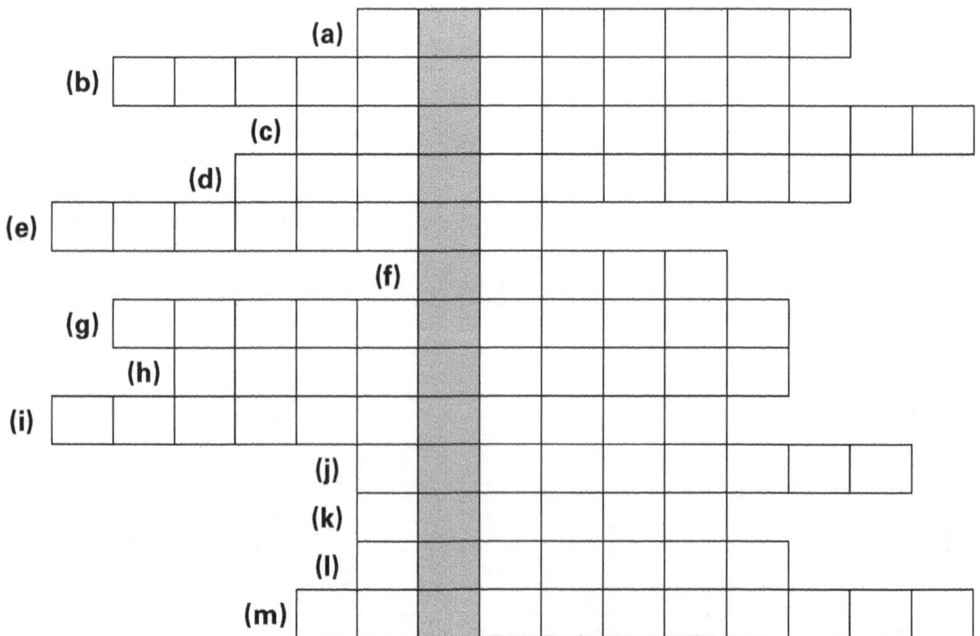

2 Write the word that you have created in the shaded column, along with its definition.

____

RATE MY UNDERSTANDING
Shade the face that shows your rating

7.12 Thinking about my learning

1 This chapter involved many new terms and several cycles which needed to be understood and remembered. Think about how you went about committing this information to your memory.

(a) What strategies did you use? ______________________________

____

____

(b) Which of these strategies seemed to work well?

____

____

(c) Which strategies that you used were less successful?

____

____

(d) What do you think made the difference between the successful and unsuccessful strategies? ______________________________

____

____

2 Talk to at least four other members of the class about how they study and remember important information.

(a) List any study techniques that your class members use and have found successful. Focus on learning about strategies and techniques which are different to your own. You should aim to find at least two new methods.

____

____

____

(b) What advantages and disadvantages can you see with these methods?

____

____

3 Having considered your own successful learning strategies and those of your class members, what do you think are the most important elements of a successful strategy for learning?

____

____

____

CHAPTER 8

The Universe

8.1 Knowledge preview

Science understanding

FOUNDATION	STANDARD	ADVANCED

1 Create a concept map using as many of the following terms as you can.

big bang	black hole	dwarf	fusion	galaxy	gas cloud
gas giant	light year	magnitude	main sequence	Milky Way	nebula
parsec	planet	red giant	star	supernova	universe

2 Select any four of the terms from question 1. Write an explanation of each term.

3 Write down three questions you have about the universe that you have often wondered about.

8.2 The Titius–Bode law

Science inquiry skills

FOUNDATION | STANDARD | **ADVANCED**

Processing & Analysing

Throughout history, cosmologists have been fascinated by the power of mathematics to predict and explain aspects of the universe. The Titius–Bode law is a mathematical rule that can be used to estimate the average distance between the Sun and many of the planets and dwarf planets of the solar system. It was formulated by Johann Titius in 1766 and published by Johann Bode in 1772. (Unfairly, Bode did not give any credit to Titius for his discovery.)

astronomical unit (AU) (*n*) a unit of length based on the average distance between the Earth and the Sun

coincidence (*n*) events that happen at the same time by chance

correspond to (*v*) to match, to be equal to

cosmologist (*n*) a scientist who studies the universe

dwarf (*adj*) small

successive factors of two (*n*) 2, 4, 6, 8, 16 etc.

The Titius–Bode law

The law is usually written as: $a = 0.4 + 0.3 \times k$

where a is the distance between a planet and the Sun in astronomical units (AU) and $k = 0, 1, 2, 4, 8, 16$ and so on up through successive factors of two.

Examples using the law

Mercury, the closest planet to the Sun, corresponds to $k = 0$:

$a = 0.4 + 0.3 \times 0 = 0.4$ AU This is the average distance between Mercury and the Sun.

The next planet from the Sun, Venus, corresponds to $k = 1$:

$a = 0.4 + 0.3 \times 1 = 0.7$ AU This is the average distance between the Sun and Venus.

Earth corresponds to $k = 2$ and Mars to $k = 4$.

Using a value of $k = 8$ it is possible to predict the distance to Ceres, the dwarf planet that is part of the asteroid belt.

At the time when Titius formulated the law and Bode published it, only the inner planets to Saturn had been discovered. Bode was particularly excited by William Herschel's discovery of Uranus in 1781 as its position fitted the Titius–Bode law very well. (In fact, it was Bode who suggested the name 'Uranus' for the planet; Herschel preferred to call it 'Georgium Sidus' in honour of King George III.)

While the match between the law's predictions and the actual distances is not perfect, the results are very close, as shown in Table 8.2.1 on page 122.

Strangely, there is no known theoretical reason why the Titius–Bode law should work. Some scientists believe though that the match is too good to simply be coincidence. Some suggest that the law is evidence of some as yet undiscovered process in the formation of the solar system.

8.2 The Titius–Bode law

(1) Use the Titius–Bode law to complete Table 8.2.1. Show your calculations in the space below.

Table 8.2.1 Results of the Titius–Bode law and actual distances between the Sun and the planets

Planet	k	Titius–Bode law distance (AU)	Real distance (AU)
Mercury	0	0.4	0.39
Venus	1	0.7	0.72
Earth	2	1.0	1.00
Mars	4	1.6	1.52
Ceres (dwarf planet)	8		2.77
Jupiter	16		5.20
Saturn	32		9.54
Uranus	64		19.2
Neptune			30.06
Pluto (dwarf planet)			39.44

(2) Note that the law does not work well for Neptune. Which body's orbit is predicted more accurately by $k = 128$?

(3) **(a)** Which planet (other than Neptune and Pluto) has the biggest difference between the predicted and real value?

(b) What is this difference?

4 Discuss whether or not you think the accuracy of the Titius–Bode law is just a coincidence or the result of an as yet unknown cosmic law.

RATE MY UNDERSTANDING
Shade the face that shows your rating

8.3 Naming stars

Science understanding

FOUNDATION | STANDARD | **ADVANCED**

Only a few stars are well known enough to have their own names. Most stars are named according to the constellation in which they appear.

Figure 8.3.1 shows the five stars of the Southern Cross. In astronomy, the Southern Cross is known as Crux (Latin for cross). The brightest star is the one at the bottom of the cross; it is called α-Crucis. The dimmest is called ε-Crucis because it is fifth brightest. The two pointer stars on the left, α-Centauri and β-Centauri, are not part of the Southern Cross constellation.

apparent magnitude (*n*) a measure of how bright a star appears from Earth

astronomical (*adj*) relating to space and the stars

constellation (*n*) a formation of stars that may look like an object or animal

designation (*n*) naming system

exception (*n*) something that does not follow the rule

Figure 8.3.1 The Pointers (α-Centauri and β-Centauri) and the Southern Cross (Crux)

Within a constellation, each star is generally named according to its apparent magnitude. The brightest star in a constellation is always designated alpha (α), the first letter of the Greek alphabet (shown in Table 8.3.1 on page 124). The other stars in the constellation are then named beta (β), gamma (γ), delta (δ) and so on in order of their decreasing brightness.

Star naming

Consider the Centaurus constellation, the constellation next to the Southern Cross. The astronomical name of Alpha Centauri (or α-Centauri) comes from the fact that it is the brightest star in the constellation Centaurus. The other Pointer star is called β-Centauri because it is the next brightest star in that constellation. Notice that the name of the constellation is often slightly modified when used to name a star. Instead of Centaurus, the stars are known as Centauri. Similarly, Crux (for Southern Cross) is modified to Crucis.

This system of naming stars according to their constellation and brightness was invented by the German scientist Johann Bayer in the 17th century and is known as the Bayer designation.

There are some exceptions to the rule that the brightest star is given the first letter of the Greek alphabet. Where two stars are similar in brightness, the one that rises in the east first is named alpha.

This slightly confusing system was developed because in the 17th century astronomers did not have instruments that could allow them to determine the exact brightness of stars.

8.3 Naming stars

Exceptions when naming stars

Castor and Pollux are the two brightest stars in the constellation Gemini. They are of almost equal brightness, but Pollux is the brighter of the two. For this reason, Pollux would normally be designated as Alpha Geminorum. However, Pollux is instead designated as Beta Geminorum because it rises after Castor.

Table 8.3.1 Letters of the Greek alphabet

Letter (lower case)	Name
α	alpha
β	beta
γ	gamma
δ	delta
ε	epsilon
ζ	zeta
η	eta
θ	theta
ι	lota
κ	kappa
λ	lambda
μ	mu
ν	nu
ξ	xi
ο	omicron
π	pi
ρ	rho
σ	sigma
τ	tau
υ	upsilon
φ	phi
χ	chi
ψ	psi
ω	omega

1. The two brightest stars in the constellation Lyra are commonly known as Vega and Sheliak. Vega is also known as α-Lyrae and Sheliak is known as β-Lyrae.

   (a) Which star will be brighter when viewed from Earth?

   ____

   (b) Explain your answer.

   ____

   ____

2. The two brightest stars in the constellation Cygnus (the Swan) are called Deneb (brightest) and Albireo (second brightest). Use the Bayer designation for naming stars in a constellation to identify which one would be called Alpha Cygni and which would be Beta Cygni.

   ____

   ____

3. Table 8.3.2 contains the names and Bayer designations of stars in the constellation Orion. Use the Bayer designation to reorder the following list of stars on Table 8.3.2, in order of decreasing brightness from 1 to 7, where 1 is the brightest and 7 is the least bright.

Table 8.3.2 Stars in the Orion constellation

Common name	Bayer designation	Brightness (1—brightest, 7—least bright
Alnilam	ε-Orionis	
Alnitak	ζ-Orionis	
Bellatrix	γ-Orionis	
Betelgeuse	α-Orionis	
Mintaka	δ-Orionis	
Rigel	β-Orionis	
Saiph	κ-Orionis	

4. In the TV show *Futurama*, a group of aliens come from a planet called Omicron Persei 8. This is the eighth planet orbiting a star in the constellation Perseus.

   (a) Use the Bayer designation to calculate how many stars in this constellation must appear brighter than Omicron Persei when observed from Earth.

   ____

   (b) Explain why it is unlikely that the aliens from this planet would call their planet's sun Omicron Persei.

   ____

   ____

RATE MY UNDERSTANDING
Shade the face that shows your rating

8.4 The Hertzsprung–Russell diagram

Science inquiry skills

FOUNDATION | **STANDARD** | ADVANCED

Processing & Analysing | Communicating

The Hertzsprung–Russell (or H–R) diagram plots the temperature of a star on the horizontal axis of the chart and absolute magnitude (brightness) on the vertical axis. Absolute magnitude is a measure of the brightness that stars would have if they were all the same distance from Earth. Table 8.4.1 gives values for over 50 stars. Our Sun's values are highlighted in the shaded row.

Table 8.4.1 Star temperature and magniture (brightness)

Star	Temperature (K)	Absolute magnitude
40 Eridani A	4900	6.0
40 Eridani B	10 000	11.1
61 Cygni A	4130	7.6
61 Cygni B	3870	8.4
70 Ophiuchi A	4950	5.8
70 Ophiuchi B	3870	7.5
Achemar	20 500	−2.4
Acrux	28 000	−4.0
Adhara	23 000	−5.2
Al Na'ir	15 550	−1.1
Aldebaran	4130	−0.8
Alhena	9900	0.0
Alioth	9900	0.4
Alkaid	20 500	−1.7
Alnilam	26 950	−6.2
Alnitak	33 600	−5.9
Altair	8060	2.2
Arcturus	4590	−0.4
Atria	4590	−0.1
Avior	4900	−2.1
Bellatrix	23 000	−4.3
Canopus	7400	−3.1
Capella	5150	−0.6
Castor	9620	1.2
Delta Canis Majoris	6100	−8.0
Alpha Centauri	5840	4.3
Beta Centauri	25 500	−5.1
Alpha Crucis	20 500	−3.3
Beta Crucis	28 000	−4.7

Star	Temperature (K)	Absolute magnitude
Dubhe	4900	0.2
Elnath	12 400	−1.6
Epsilon Eridani	4590	6.1
Epsilon Indi	4130	7.0
Fomalhaut	9060	2.0
Gacrux	3750	−0.5
Hadar	25 500	−5.3
Kaus Australis	11 000	−0.3
Menkalinan	9340	0.6
Miaplacidus	9300	−0.6
Mirfak	7700	−4.6
Mirzam	25 500	−4.8
Peacock	20 500	−2.3
Polaris	6100	−4.6
Pollux	4900	1.0
Procyon A	6580	2.6
Procyon B	9700	13.0
Regulus	13 260	−0.8
Rigel	12 140	−7.2
Shaula	25 500	−3.4
Sirius A	9620	1.4
Sirius B	14 800	11.2
Spica	25 500	−3.4
Sun	**5840**	**4.8**
Tau Ceti	5150	5.7
Theta Scorpii	7400	−5.6
van Maanen's Star	13 000	14.2
Vega	9900	0.5

8.4 The Hertzsprung–Russell diagram

1. Construct your own H–R diagram by plotting the stars from Table 8.4.1 on page 125 onto the grid below.

   Note that the horizontal axis is drawn to a logarithmic scale. This means that the values of grid lines are not evenly spaced as in a normal linear scale. (In fact, each grid line is a factor of $10^{0.1}$ or 1.26 times higher than the previous one.)

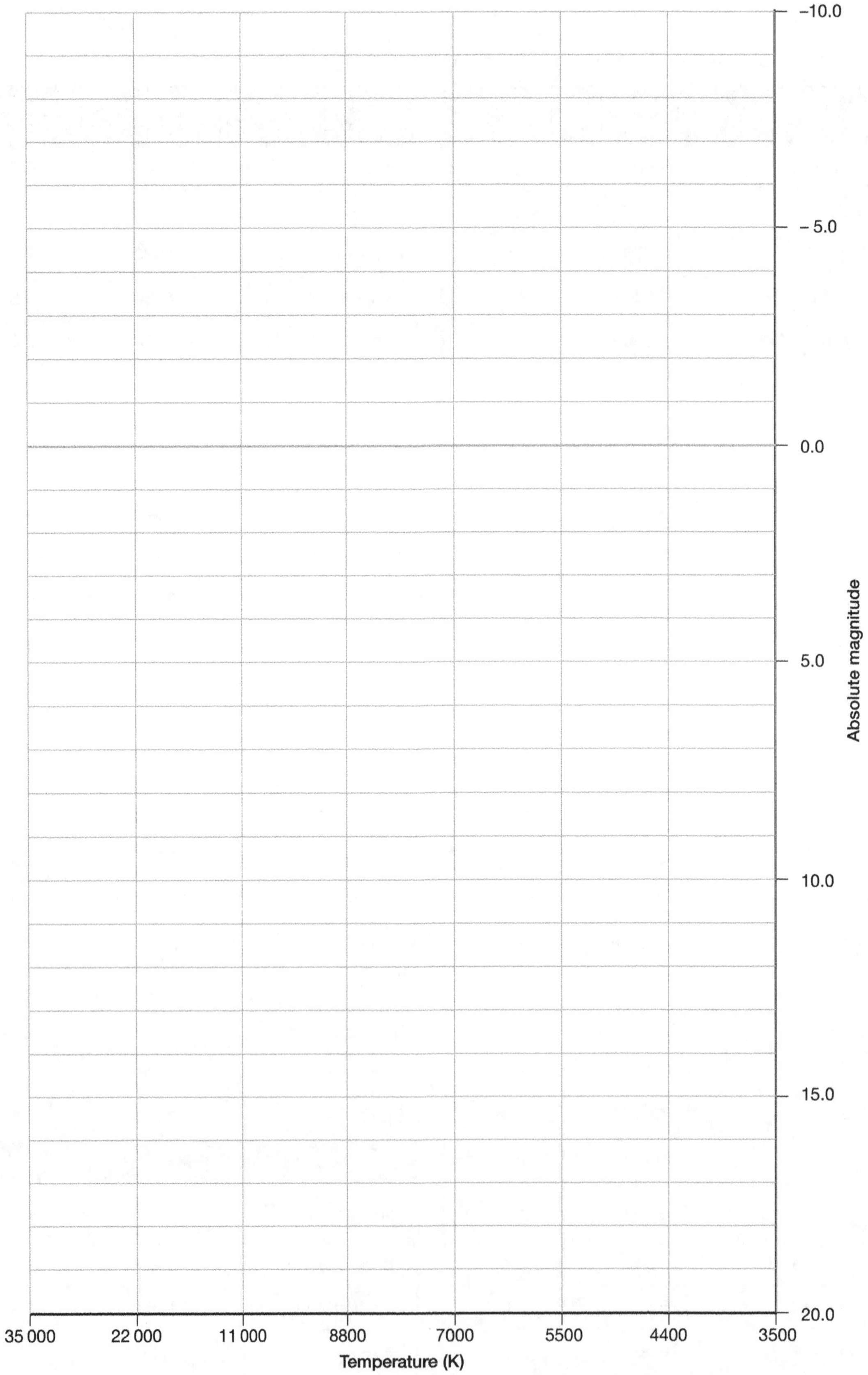

Absolute magnitude
–10.0
–5.0
0.0
5.0
10.0
15.0
20.0

Temperature (K): 35 000, 22 000, 11 000, 8800, 7000, 5500, 4400, 3500

RATE MY UNDERSTANDING
Shade the face that shows your rating   

8.5 History of cosmology

Science understanding

FOUNDATION | STANDARD | **ADVANCED**

Cosmology is the study of everything in the universe. Table 8.5.1 lists some of the major events in the history that changed our understanding of the structure of the universe.

Table 8.5.1 Major historical events in universe understandings

Date	Event
around 350 BCE	Aristotle describes the geocentric (Earth-centred) model of the universe.
around 200 BCE	Aristarchus proposes a heliocentric (Sun-centred) model of the universe that is widely ignored.
around 150 BCE	Ptolemy refines the geocentric model of the universe to try to explain the retrograde (east to west) motion of the planets.
1520	Ferdinand Magellan is the first European to observe the Magellanic Clouds, which are later recognised as galaxies outside the Milky Way.
1543	Nicholas Copernicus publishes a paper that presents the heliocentric model of the universe with a number of improvements on Aristarchus' version.
1600	Giordano Bruno is burnt at the stake. One of his 'crimes' was promoting the heliocentric model of the universe.
1609	Johannes Kepler publishes *Astronomia Nova*, explaining the motion of the planets by using a heliocentric model of the universe and elliptical (oval-shaped) orbits.
1610	Galileo Galilei observes the moons of Jupiter.
1632	Galileo publishes his *Dialogue on Two World Systems*, which argues in favour of the heliocentric model of the universe.
1687	Isaac Newton publishes *Principia Mathematica* in which, among other things, he explains the law of universal gravitation (gravity).
1750	Thomas Wright suggests that the Milky Way galaxy is a flattened disk of stars and that nebulae might be other galaxies.
1781	William Herschel discovers Uranus, the first planet not visible to the naked eye to be discovered.
1785	William Herschel constructs the first map of the Milky Way galaxy.
1838	Friedrich Bassel observes stellar parallax—the apparent movement of stars due to the Earth's movement around the Sun.
1846	John Couch Adams discovers Neptune.
1922	Ernst Opik demonstrates that the Andromeda nebula lies outside the Milky Way galaxy.
1927	Georges Lemaitre proposes the big bang theory, which suggests that the universe began as a giant explosion of energy.
1929	Edwin Hubble discovers evidence of an expanding universe, which supports the big bang model.
1930	Clyde Tombaugh discovers Pluto.
1964	Arno Penzias and Robert Wilson discover cosmic microwave background radiation, further evidence supporting the big bang model.
1969	Neil Armstrong and Buzz Aldrin are the first human beings to walk on the surface of the Moon.
1983	*Pioneer 10* becomes the first spacecraft to leave the solar system.
1998	The spacecraft *Galileo* returns data suggesting that Europa has liquid oceans under its icy surface.
2005	Planet-sized objects Eris, Haumea and Makemake are discovered in the Kuiper Belt.
2006	The International Astronomical Union rules that Pluto is not a planet.
2011	Brian Schmidt and Saul Perlmutter win the Nobel Prize for discovering that the rate of expansion of the universe is increasing.

8.5 History of cosmology

1 Using the information presented in Table 8.5.1 on page 127, choose the 10 events that you consider most significant to developing our understanding of the cosmos and construct a timeline depicting these on the blank timeline below.

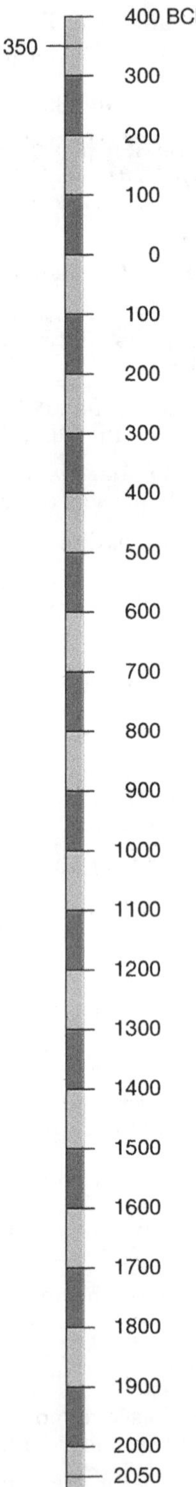

2 Justify two of your choices.

RATE MY UNDERSTANDING
Shade the face that shows your rating

8.6 The Hubble constant

Science inquiry skills

FOUNDATION | **STANDARD** | ADVANCED

Communicating

In 1929, Edwin Hubble published his research into the motion of galaxies. This research fundamentally changed our understanding of the cosmos as it provided the first evidence that the universe was expanding.

For each of the galaxies he studied, Hubble measured two properties—its distance from the Milky Way galaxy in megaparsecs (Mpc) and its recession velocity, that is, the speed with which it is moving away from our galaxy.

galaxy (*n*) a large system of stars

megaparsec (*n*) a unit used to measure very large distances in space

The data collected by Hubble is shown in the following table.

Table 8.6.1 Hubble's motion of galaxies data

Distance (Mpc)	Recession velocity (km/s)
0.032	170
0.034	290
0.214	–130
0.263	–70
0.275	–185
0.275	–220
0.45	200
0.5	290
0.5	270
0.68	200
0.8	300
0.9	–30
0.9	650
0.9	150
0.9	500
1	920
1.1	450
1.1	500
1.4	500
1.7	960
2	500
2	850
2	800
2	1090

(1) Construct a graph for this data on the grid on the following page.

8.6 The Hubble constant

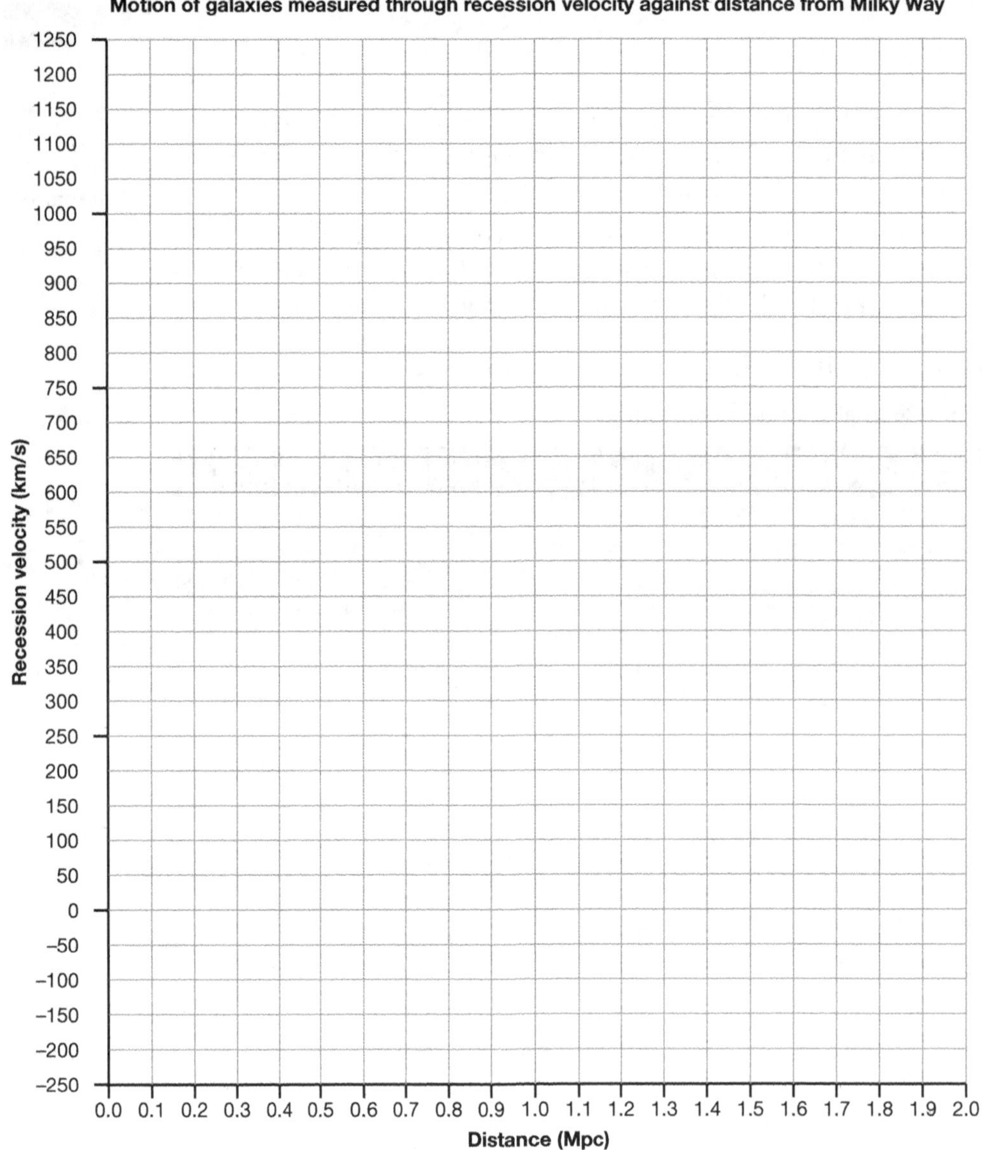

(2) Draw a line of best fit through the points. (Note: Hubble's techniques have been refined in recent years to provide a much smaller spread of data.)

(3) Calculate the gradient of the line of best fit. This is known as the Hubble constant, H_0. The units of the Hubble constant are km/s/Mpc. Hubble measured this value as being about 500 km/s/Mpc. Gradient is calculated as rise divided by run.

____

____

4 Notice that at the bottom left-hand corner of the graph, there are a number of galaxies that have negative recession velocities. This means that they are moving towards our galaxy rather than away from it. This contradicts the idea of an expanding universe. Propose a reason why galaxies that are close to the Milky Way might be moving towards it rather than away from it.

____

____

RATE MY UNDERSTANDING
Shade the face that shows your rating   

8.7 Literacy review

Science understanding

FOUNDATION	STANDARD	ADVANCED

1 The table lists some terms and definitions about the Universe. Fill in the missing sections to complete the table.

Term	Definition
	the distance light travels in a year
	theory that the universe began with an enormous explosion of energy
neutron star	
	measure of the brightness of a star
	collapsing cloud of gas that will eventually become a star
accretion	
	stretching of light waves due to the motion of stars away from Earth
red giant	
cosmology	
	astronomical unit of length equivalent to 3.26 light- years
parallax	
	the galaxy in which our solar system is located
white dwarf	
singularity	
	giant explosion that occurs when a star runs out of nuclear fuel

2 Read each of the following statements and circle whether they are True or False according to current theories of cosmology.

Statement	True / False
Gravity is a force that causes matter to come together and is measured in parsec.	True / False
The process of nuclear fusion involves the conversion of helium into hydrogen to make light and heat.	True / False
A black hole can distort space so that light rays bend, to cause gravitational lensing.	True / False
A nebula is a cloud of gas and is the place where new stars are born.	True / False
At the centre of a galaxy there is a supermassive black hole.	True / False
The Sun is a white dwarf.	True / False
A red giant is younger than a white dwarf and a black dwarf is older than both of these.	True / False
The steady state theory is currently the most widely supported theory about the history of the universe by scientists..	True / False
Two stars that orbit a common mass are called twin stars.	True / False
Black holes have strong gravitational fields.	True / False

RATE MY UNDERSTANDING
Shade the face that shows your rating

8.8 Thinking about my learning

1 At the beginning of this unit you prepared a concept map using some terms. The same terms are listed below. Use these terms and any others which you think apply to this unit to create a new concept map.

big bang	black hole	dwarf	fusion	galaxy	gas cloud
gas giant	light year	magnitude	main sequence	Milky Way	nebula
parsec	planet	red giant	star	supernova	universe

2 Compare your new concept map with the one you prepared on page 120. How has your understanding of the universe changed?

3 (a) Write down the three most important things that you have learned.

(b) Compare your list with other members of the class. Do others in the class have the same things on their list? If they don't, why do you think you have different ideas about what is important?

CHAPTER 9

Motion and energy

9.1 Knowledge preview

Science understanding

FOUNDATION | STANDARD | ADVANCED

1 Fill in the table with the correct terms from the box.

distance energy energy efficiency force gravitational potential energy inertia kinetic energy mass speed time

Concept	Definition
	a measurement of how far apart objects are
	the rate of change of distance (how fast an object is moving)
	how long it takes to do something
	a measure of how much matter there is in an object
	a pull or push
	this is measured in joules and is needed to make anything happen
	the energy of a moving body
	the tendency of an object to resist changes in motion
	energy stored in an object due to its position
	the ratio of useful energy output compared to total energy input

2 (a) List some of the different types of forces that you know. ______________

(b) List three things that forces can do. ______________

(c) Explain what happens to an object when the forces are balanced. ______________

9.2 Driver distractions

Science understanding

FOUNDATION | **STANDARD** | ADVANCED

Read the following article about the effect of mobile phone use in cars and then answer the questions that follow.

distract (*v*) to take someone's attention away from what they are doing

impaired (*adj*) restricted, made worst

manoeuvre (*n*) action, movement

navigate (*v*) to find the way

simulator (*n*) a machine that mimics driving a car, used to test drivers

Mobile phones distract drivers more than passengers do

Mobile phone calls distract drivers far more than even the chattiest passenger, causing drivers to follow too closely and miss exits, US researchers report.

Using a hands-free device does not make things better and the researchers believe they know why—passengers act as a second set of eyes, shutting up or sometimes even helping when they see the driver needs to make a maneuver.

The research, published in the *Journal of Experimental Psychology: Applied*, adds to a growing body of evidence that mobile phones can make driving dangerous.

Dave Strayer of the University of Utah and colleagues have found in a series of experiments using driving simulators that hands-free mobile phones are just as distracting as hand-held models.

They have demonstrated that chatting on a mobile phone can slow the reaction times of young adult drivers to levels seen among senior citizens, and shown that drivers using mobile telephones are as impaired as drivers who are legally drunk.

Figure 9.2.1 Drivers using a hands-free device drift out of their lanes and miss exits more frequently, say researchers.

For their latest study, also using a simulator, Strayer's team showed that drivers using a hands-free device drifted out of their lanes and missed exits more frequently than drivers talking to a passenger. They tested 96 adults aged 18 to 49.

'The passenger adds a second set of eyes, and helps the driver navigate and reminds them where to go,' Strayer said in a statement.

'When you take a look at the data, it turns out that a driver conversing with a passenger is not as impaired as a driver talking on a cell phone,' he added.

Passengers also simplify conversation when driving conditions change, the researchers wrote.

'The difference between a cell phone conversation and passenger conversation is due to the fact that the passenger is in the vehicle and knows what the traffic conditions are like, and they help the driver by reminding them of where to take an exit and pointing out hazards,' Strayer said.

Source: Reuters, December 2008

9.2 Driver distractions

1 Researchers have found that using a hands-free mobile device is more distracting than having a conversation with a passenger in the car.

(a) Describe how using a mobile phone affects a driver's reaction time.

(b) State the two comparisons made by the researchers to describe the effect of this altered reaction time.

2 Name two driving behaviours that the researchers found were more likely to occur when a driver was speaking on a mobile phone.

3 Explain why speaking to a person in a car is not as distracting as having a conversation on a mobile phone while driving.

4 List arguments for and against using a hands-free mobile device while driving.

Advantages of using a hands-free mobile device while driving	Disadvantages of using a hands-free mobile device while driving

5 **(a)** Do you think there are any situations when drivers should be allowed to use a hands-free mobile device while in a car?

(b) Justify your answer.

9.3 Road statistics

Science understanding

FOUNDATION | **STANDARD** | ADVANCED

The data presented in the following questions is taken from the 'Road Deaths Australia 2014 Statistical Summary' compiled by the Australian Government Department of Infrastructure, Transport, Regional Development and Local Government.

1. Figure 9.3.1 shows the total number of people who died on Australian roads each year from 1981 to 2014.

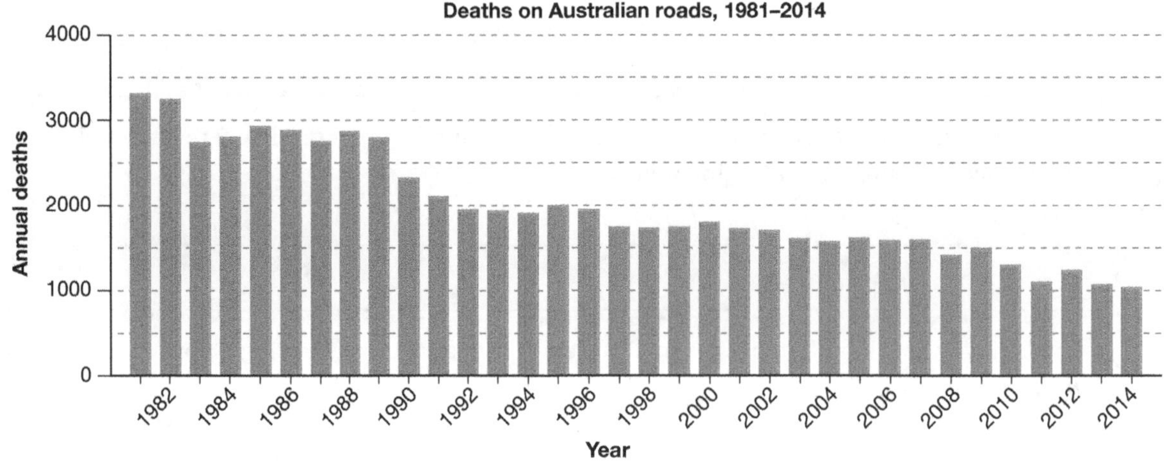

Figure 9.3.1 Australian annual road deaths

(a) Describe the trend in the numbers of people killed on Australian roads over this period.

(b) Suggest reasons to explain the trend.

2. Figure 9.3.2 illustrates the breakdown of deaths per road-user group according to gender in 2008.

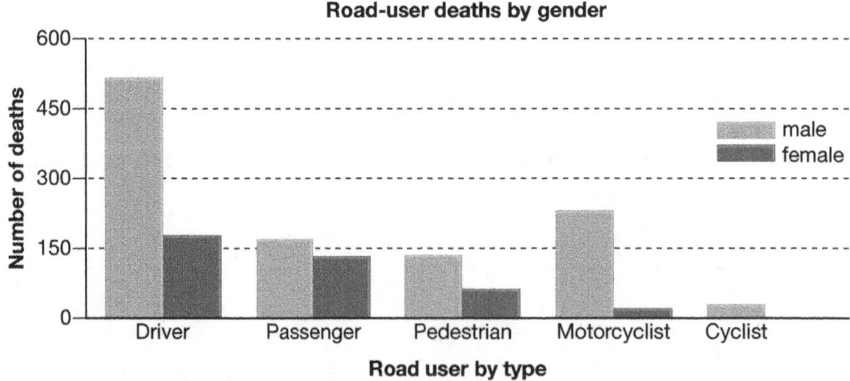

Figure 9.3.2 Death by road-user group and gender in 2008

(a) State the type of road user (driver, passenger, pedestrian, motorcyclist or cyclist) that was most often killed in an accident.

(b) Suggest a reason to explain this finding.

(c) In 2008, 516 of drivers killed in car accidents were male and 178 were female. In 2007, 615 of those killed were male and 168 were female. Suggest reasons to explain this difference in numbers between males and females.

(d) Explain why many more motorcyclist deaths may have occurred among males.

3 Figure 9.3.3 provides data about the speed limit at crash sites.

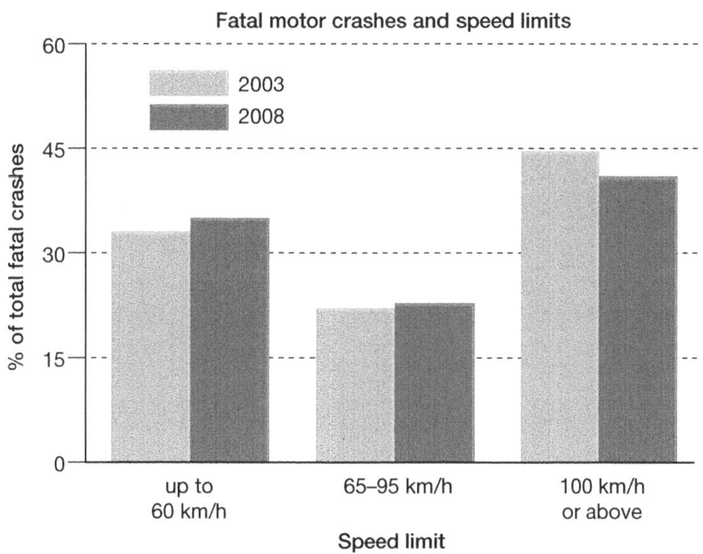

Figure 9.3.3 Speed limit at crash site

(a) State the speed limit zone in which the highest number of road deaths occurred.

(b) Explain why you think this is this case.

4 Study the three graphs shown in Figure 9.3.4. They show the time accidents occurred, where a pedestrian or driver was killed, and the ages of the deceased.

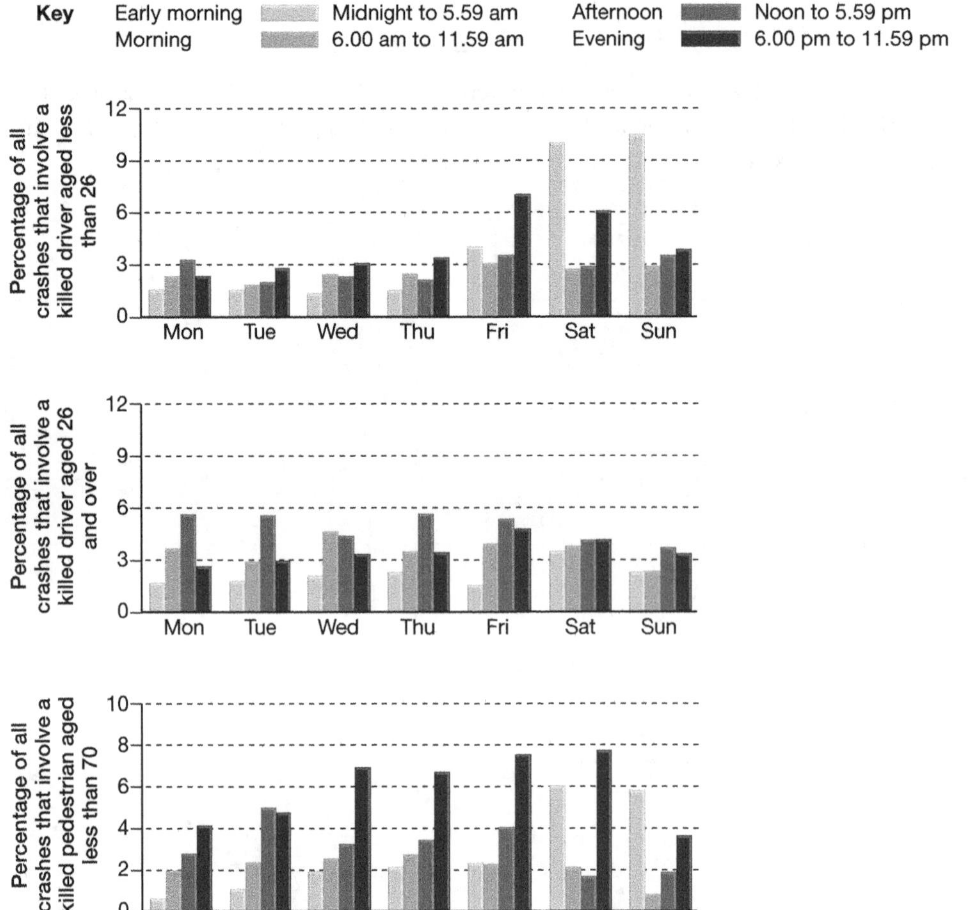

Figure 9.3.4 Crashes involving death of selected road users by time of day and day of week

(a) State the time of day and days of the week when drivers under the age of 26 were at a much greater risk of having an accident.

(b) State the time of day that generally involved the most accidents for drivers over the age of 26.

(c) Propose reasons why your responses to the last two questions were different.

(d) State the time of day that pedestrians were at greatest risk of being hit by a car.

(e) Propose why the greatest number of accidents involving pedestrians happened in this time interval.

RATE MY UNDERSTANDING
Shade the face that shows your rating   

9.4 Speed–time graphs

Science inquiry skills

FOUNDATION | **STANDARD** | ADVANCED

Processing & Analysing

1. The table below shows the instantaneous speed recorded on the speedometer in a car at every minute of a 15-minute journey.

Time (min)	0	1	2	3	4	5	6	7	8	9	10	11	12	13	14	15
Speed (km/h)	0	22	47	50	50	50	35	11	29	60	32	3	0	0	25	60

(a) Use this data to construct a speed–time graph of the car's motion on the axes below.

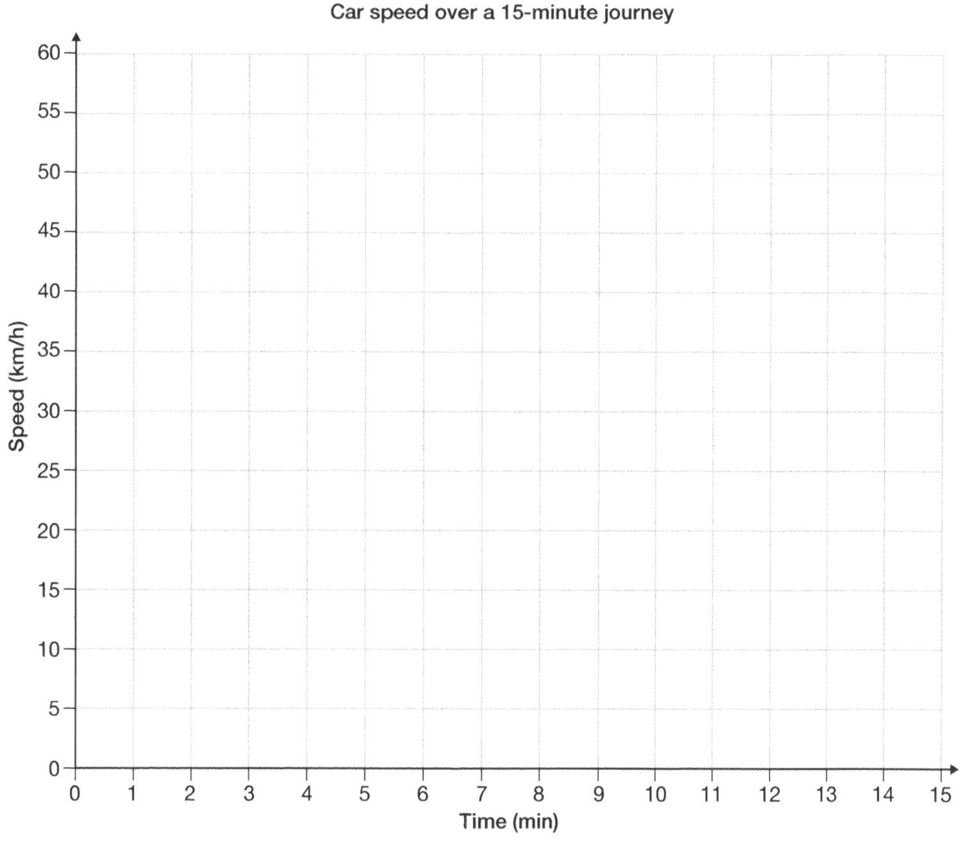

(b) (i) Using a red pen, identify on your graph when the car was speeding up.

(ii) Using a blue pen, identify on your graph when the car was slowing down.

(c) (i) At one stage of the journey, the driver slowed down and turned a corner. Propose how many minutes into the trip this occurred.

____

(ii) At another stage of the journey, the driver stopped at a set of traffic lights. Propose how many minutes into the trip this occurred.

____

2. A family set out on a long car journey to reach a holiday destination. A speed–time graph of their motion is shown below. They drive for a certain time on freeways before having a break for a meal, then drive a further distance through a major city before reaching their accommodation.

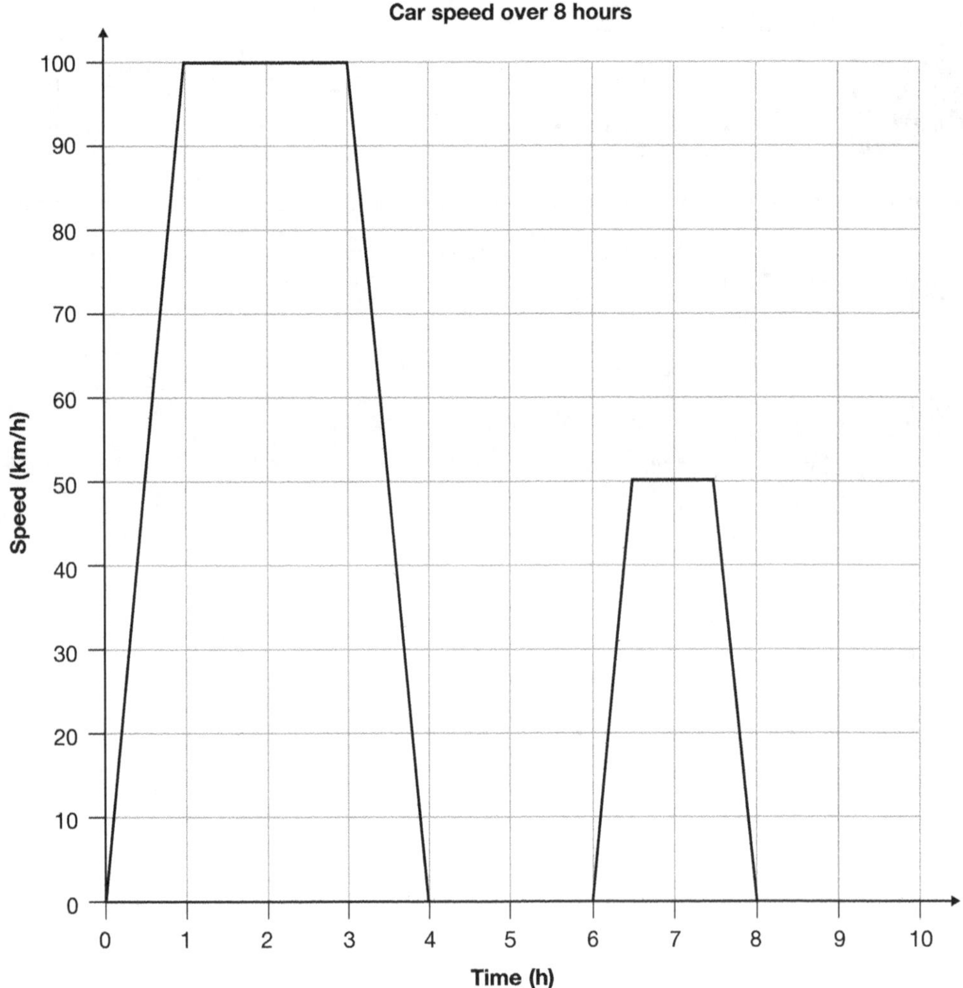

(a) Describe the family's motion:

(i) 1–3 hours into the trip ______

(ii) 3–4 hours into the trip ______

(iii) 4–6 hours into the trip ______

(iv) 6–6.5 hours into the trip ______

(v) 6.5–7.5 hours into the trip. ______

(b) The distance travelled is equivalent to the area below the speed–time graph for the journey. Use this fact to calculate the distance the family travelled to reach their holiday destination.

______

______

RATE MY UNDERSTANDING
Shade the face that shows your rating

9.5 Distance– and displacement–time graphs

Science understanding

FOUNDATION | STANDARD | **ADVANCED**

(1) Jake, Charlie and Nasir compete in a 100-metre race. The distance–time graph of their motion is shown. Analyse this graph to answer the following questions.

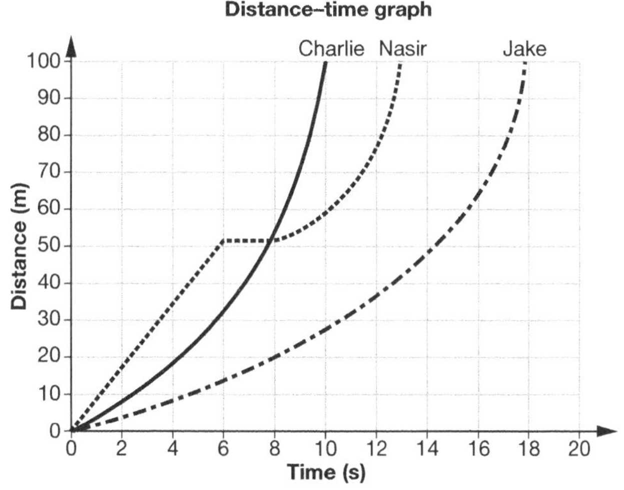

(a) State who won the race. ____________________

(b) Calculate the average speed of the three runners in metres per second (m/s). Express your answer to one decimal place.

(i) Charlie's average speed = $\frac{\text{distance travelled}}{\text{time taken}}$ = ____________________

(ii) Jake's average speed = ____________________

(iii) Nasir's average speed = ____________________

(c) Compare Nasir's motion to those of Charlie and Jake.

____

____

____

(2) The distance–time graph shown below plots Lucy's motion as she rides her bike along the bike path at the beach. Analyse Lucy's motion to answer the following questions:

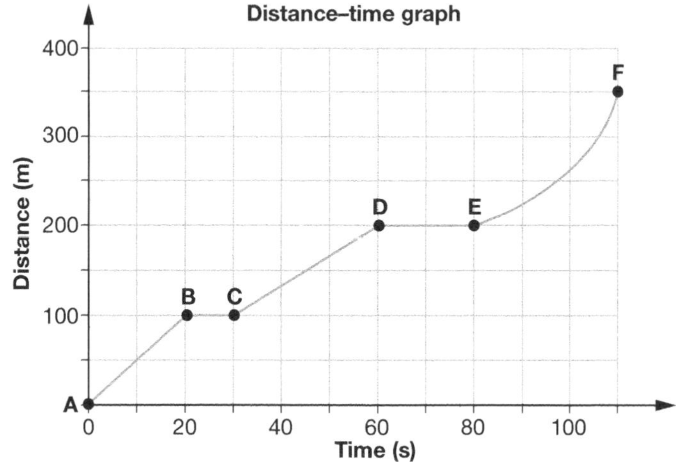

(a) Calculate Lucy's average speed as she travels from A to B.

____

(b) Calculate Lucy's average speed as she travels from C to D.

____

(c) Identify the interval(s) in which Lucy:

(i) stops ______________________________

(ii) travels with constant speed ______________________________

(iii) travels with increasing speed. ______________________________

(d) Calculate Lucy's average speed for the entire journey.

____

(e) Explain how this average speed does not fully describe Lucy's motion over the entire journey.

____

____

____

(f) Assuming the bike path followed a northerly direction, state Lucy's displacement for the journey.

____

3 Calculate the gradient ($\frac{\text{rise}}{\text{run}}$) for each stage of the two displacement–time graphs shown below. Use your results to plot the corresponding velocity–time graph for each on the axes provided.

(a)

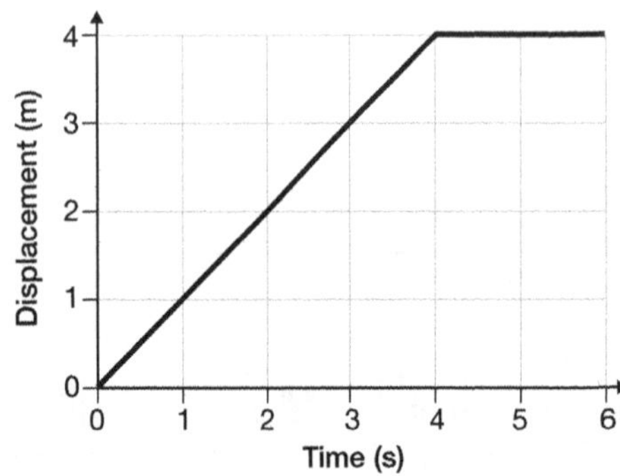

Velocity (m/s)

+2
+1
0
−1
−2

1 2 3 4 5 6

Time (s)

(b)

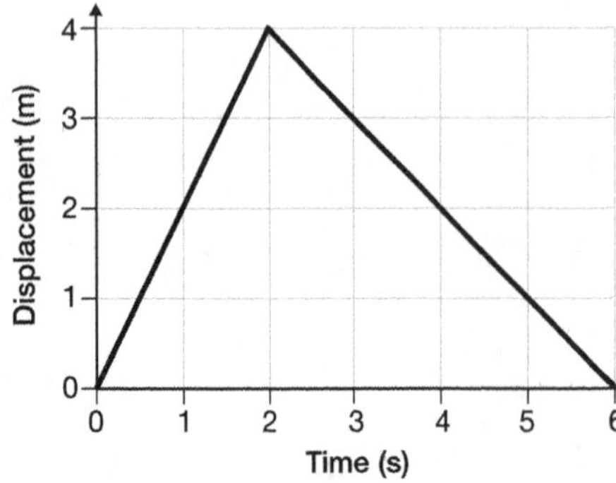

Velocity (m/s)

+2
+1
0
−1
−2

1 2 3 4 5 6

Time (s)

RATE MY UNDERSTANDING
Shade the face that shows your rating

9.6 Newton's second law

Science understanding

FOUNDATION	**STANDARD**	ADVANCED

Newton's second law of motion can be stated as:

$$F_{net} = m \times a$$

where F_{net} is the sum of all forces acting on an object in Newtons (N), m is the mass of the object in kilograms (kg) and a is the acceleration of the object (m/s²).

This equation can also be stated in terms of acceleration: $a = \frac{F}{m}$

and in terms of mass: $m = \frac{F}{a}$

1. Use the appropriate equation to calculate the missing quantities in the table below.

	Force (N)	Mass (kg)	Acceleration (m/s²)
(a)		6.1	2
(b)	145.6	45.5	
(c)		250	3
(d)	5645.1		6.2
(e)	5.2	0.8	
(f)	6750		5.4

2. Use the correct equation to complete each of the following tasks.

(a) If a 1200 kg car accelerates at 3.0 m/s², calculate the force supplied by its engine.

____

____

(b) Sarah accelerates at 8.0 m/s² as she starts running in a race. Given that her legs supply a force of 472 N, calculate Sarah's mass.

____

____

(c) The engine of a train of mass 250 000 kg, supplies a force of 1 525 000 N as it leaves a station. Ignoring the effects of friction, calculate the acceleration of the train.

____

____

3. Given that a falling stone will accelerate at 9.8 m/s², what is the size of the weight force that pulls the stone to Earth if the stone has a mass of 0.1 kg?

____

____

RATE MY UNDERSTANDING
Shade the face that shows your rating

9.7 Energy

Science understanding

FOUNDATION | **STANDARD** | ADVANCED

1. Calculate the kinetic energy of the following moving objects.

   (a) Diego is moving in a dodgem car at a speed of 8 m/s; the combined mass of Diego and the dodgem car is 325 kg.

   (b) A seagull of mass 0.5 kg is flying at a speed of 12 m/s.

   (c) Frankie is pushed in a pram at a speed of 2 m/s. She has a mass of 6.5 kg.

2. Calculate the gravitational potential energy in each of the following situations.

   (a) Sophie of mass 18 kg is sitting at the top of a 20 m climbing structure made from steel cables.

   (b) A paint tin of mass 200 g is resting on a 1.5 m high shelf in a garage. (**Hint:** Convert mass to kilograms.)

   (c) Taylah lifts her 4 kg sports bag onto a bench that is 1.2 m above the ground.

3. A car of mass 1210 kg is travelling at 15 m/s. The driver sees that the traffic light ahead has changed to red and applies the brakes, bringing the car to a stop.

   (a) Calculate the initial and final kinetic energy of the car.

   (b) State the change in kinetic energy as the car comes to a stop.

(4) This diagram shows the rollercoaster that Tina and Rebecca rode on at a fun park.

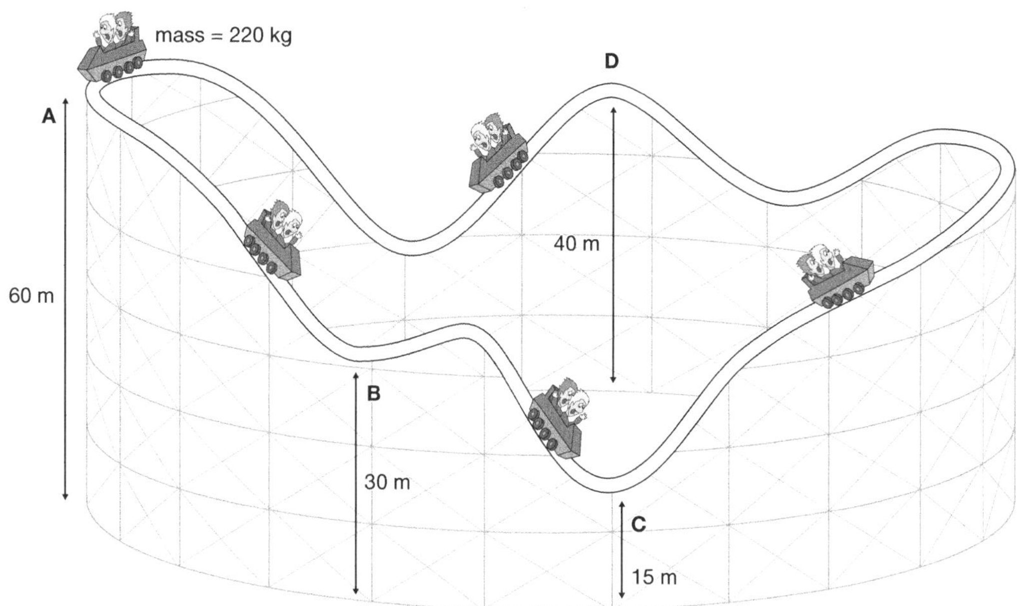

(a) Calculate their gravitational potential energy at point A at the top of the first hill.

____

____

(b) Describe how the gravitational potential energy of the cart has changed as the girls ride through point B on the track.

____

____

____

(c) Calculate the gravitational potential energy at point C.

____

____

(d) Ignoring energy losses due to friction, calculate the girls' speed at point C, rounding your answer to one decimal place.

____

____

____

____

____

____

RATE MY UNDERSTANDING
Shade the face that shows your rating

9.8 Literacy review

Science understanding

FOUNDATION | STANDARD | ADVANCED

1 Use the clues to identify the words.

Clue	Word
The straight line distance between the finishing and the starting point, written with a direction	d _ _ p l _ _ _ _ _ _ _
Time taken to react to an emergency (two words)	_ _ a _ _ _ _ _ _ _ m _
Friction force produced by moving air (two words)	_ i _ _ _ _ i _ _ _ _ _ _
The rate of change of distance	s _ _ _ _
Constant speed produced when air resistance balanced gravity (two words)	_ _ _ _ _ _ a _ _ _ l _ _ _ _ _
The rate of change of speed	a c _ _ _ _ _ _ _ _ _ _
Another name for Newton's first law	i _ _ _ _ _ a
The acceleration of an object depends upon the size of the force acting and the object's ______	m _ _ _
For every action force there is an equal and opposite ______ force.	r _ _ _ _ _ o n
The unit used to measure energy	j _ _ l _
The type of energy a moving object has	k _ _ _ _ _ c
The energy of an object due to its position is ______ potential energy.	g _ _ _ _ _ _ _ _ _ _ a l
The energy of a stretched or compressed spring is ______ potential energy.	e _ _ _ _ i _
A measure of the amount of useful energy produced	e _ _ _ _ _ _ _ _ y

2 Summarise the key equations used in forces and energy.

	Symbol	Formula
Average speed	v	
Average acceleration	a	
Newton's second law	F_{net}	
Kinetic energy	E_k	
Gravitational potential energy	E_p	
Efficiency	efficiency	

RATE MY UNDERSTANDING
Shade the face that shows your rating

9.9 Thinking about my learning

Complete questions 1 to 6 then reflect on your learning.

1 **(a)** Explain the difference between distance and displacement.

(b) Explain the difference between speed and acceleration.

2 **(a)** Explain how to calculate the speed of an object (you may use an equation in your response).

(b) Rearrange the speed equation to solve for distance.

3 Explain what is happening to the speed of the object in each of the following graphs where d is distance,
t is time
v is velocity:

(a)

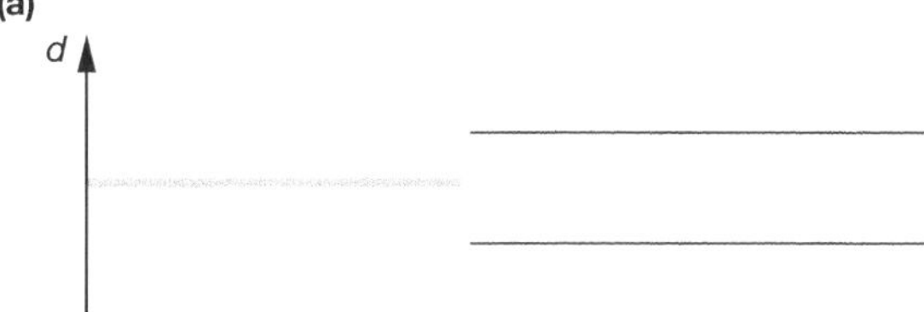

(b)

(c) **(d)**

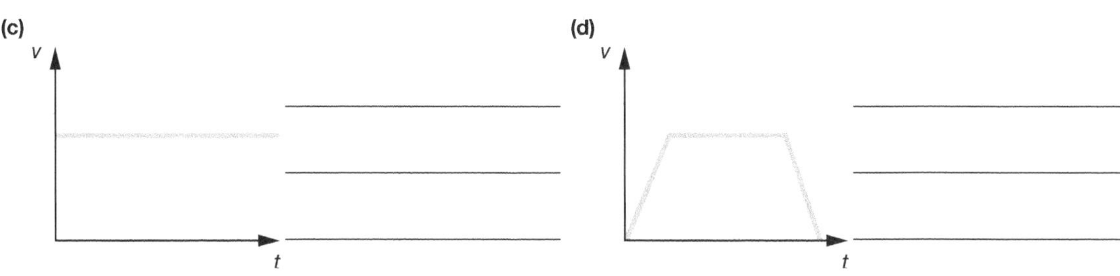

4 **(a)** Use an example to explain Newton's first law of motion.

(b) Use an example to explain Newton's second law of motion.

(c) Use an example to explain Newton's third law of motion.

5 Which has a greater effect on the kinetic energy of an object, its speed or its mass? Explain your response.

6 Explain what is meant by the term `the law of conservation of energy'.

Reflection

From your responses to the questions above, answer the following questions about your learning:

1 Which topics in this unit did you find easiest?

2 Which topics in this unit did you find the most challenging?

3 Which topics in this unit do you need to revise further?

4 What questions do you still have about the concepts in this chapter?

CHAPTER 10

Forensic science

10.1 Knowledge preview

Science understanding

FOUNDATION	STANDARD	ADVANCED

Forensic scientists use knowledge and techniques from all of the branches of science to help them to solve crimes.

The microscope is an important tool for identifying hair, fibres, soil samples and microorganisms at crime scenes.

1 Label the microscope parts using the terms in the box.

coarse focus knob	eyepiece (ocular lens)		
fine focus knob	lamp	objective lenses	stage

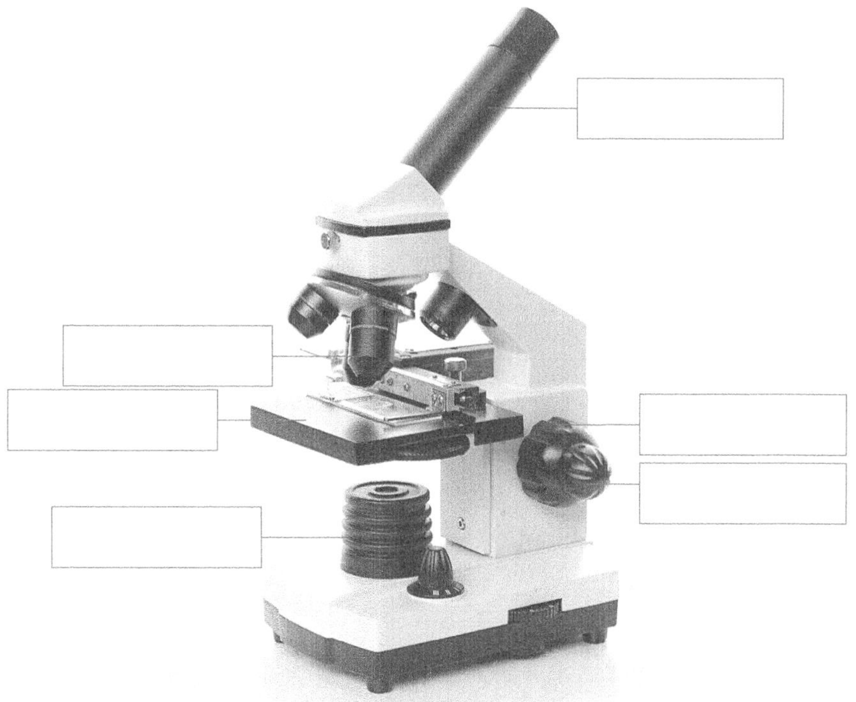

2 Genetics is important in helping to identify both criminals and victims. Describe what the term *genetics* means.

____

____

3 Explain how DNA can help identify people.

____

____

____

____

4 Separation techniques help forensic scientists identify inks used in forgeries and threatening letters, as well as helping them to identify chemicals found at crime scenes. List three separation techniques.

5 Use a diagram to help you explain how chromatography works to separate the different dyes in ink.

6 Forensic scientists use Newton's laws of motion in ballistics (the study of guns and ammunition) and to analyse blood splatters at crime scenes. What are Newton's three laws of motion?

7 Use Newton's second law of motion to explain how we can predict what direction a person was stabbed from based on the blood splatters at the crime scene.

8 Other than the examples given above, describe another scientific theory or technique that might be used in forensics.

10.2 Time of death

Science inquiry skills

FOUNDATION | STANDARD | ADVANCED

Processing & Analysing | Communicating

The human body starts cooling immediately after death. This is accompanied by other changes that can help forensic scientists tell how long a person has been dead.

Blood starts to settle or 'pool' in the lowest parts of the corpse because the heart is no longer pumping. Blood pooling usually takes around 6 hours. The location of pooling can also indicate whether the body was moved after death.

Rigor mortis is the stiffening of muscles because of a build-up of lactic acid after death. Rigor mortis moves through three phases, each lasting 6 to 12 hours. Stiffening starts in the eyelids and jaw and spreads until the whole body is stiff. The muscles then release and once again become flexible.

bloating (*n*) expanding with gas

decomposition (*n*) breaking down, rotting

goo (*n*) sticky substance

inflate (*v*) fill with gas

moult (*v*) to lose its skin or covering

Skin colour changes, first turning pale and then blue then green. The green colouring happens within 2 days and is caused by bacteria decomposing the internal tissues. Four to seven days after death, veins become more obvious, giving the skin a marble-like appearance.

Decomposition of the body proceeds in clear steps. Bloating occurs 4 to 7 days after death as decomposition gases inflate the corpse. The soft tissue of the body decays extensively in the second week, leaving little except sticky goo by the end of a month. After that, only the skeleton, nails, hair and perhaps some dried, leathery skin will be left.

Different species of insects are attracted to different stages of body decomposition and will consume flesh or lay eggs in or on the body. The eggs hatch, releasing larvae such as maggots or caterpillars. These moult and grow, each new growth stage being known as an instar. Later on, a pupa or cocoon is spun in which the larvae undergo metamorphosis, changing into their adult form. A typical life cycle is shown below in Figure 10.2.1.

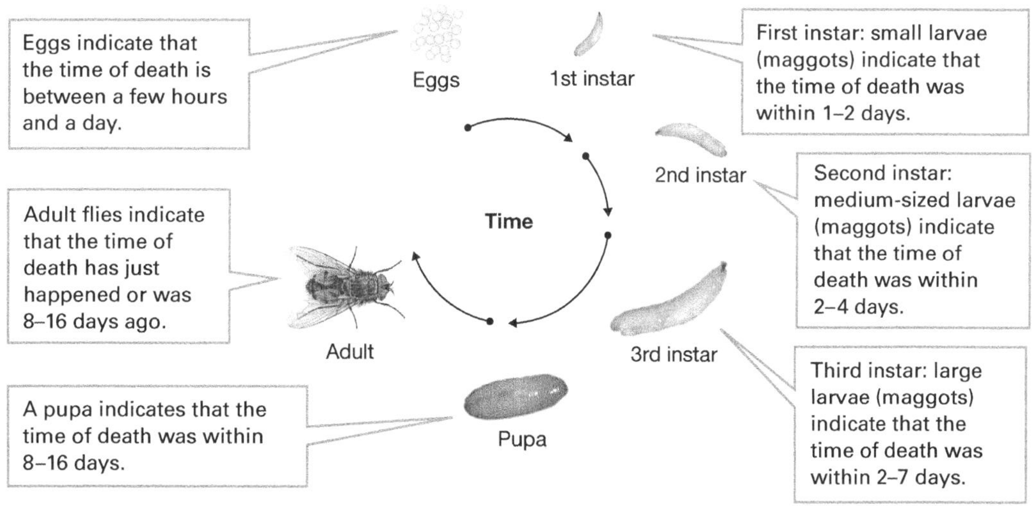

Figure 10.2.1 Life cycle of the common blowfly

1 Define the following terms:

(a) rigor mortis ______________________________

(b) bloating ______________________________

(c) larvae. ______________________________

10.2 Time of death

2 How long has a body been dead if the following are observed?

(a) the whole body is stiff ______

(b) blood is pooled ______

(c) skin is green ______

(d) the body is bloated ______

(e) large maggots are on the body ______

(f) skin is 'marbled' ______

(g) nothing is left other than a skeleton and hair ______

(h) only goo, bones and skin are left ______

(i) blowfly eggs are on the body ______

(j) eyelids are stiff but muscles are flexible. ______

3 Why doesn't blood normally pool in a living human being?

______

4 Explain how blood pooling could indicate that a body has been moved after death.

______

5 Propose a reason why skin turns pale and then blue soon after death.

______

______

6 Name the chemical that produces rigor mortis.

______

7 What causes a body to bloat?

______

8 The green colouring of bacterial decay starts near the waist. Think about the organs found around this area and propose reasons why the green colouring starts there and not elsewhere.

______

______

9 When animals die in the bush, their bodies decompose in similar stages to those of humans. Propose reasons why this is a good thing for the environment.

______

______

RATE MY UNDERSTANDING
Shade the face that shows your rating

10.3 Blood typing

Science inquiry skills

FOUNDATION | **STANDARD** | ADVANCED

Processing & Analysing | Communicating

One of the roles of blood is to rid the body of foreign bodies such as bacteria and viruses. Two chemicals are involved:

- antigens—these chemicals are attached to the outside of cells and cause your immune system to attack cells that do not belong to you
- antibodies—these chemicals are in the plasma or serum of your blood and attack any foreign particles that have entered your body.

Blood group

Blood is classified or grouped according to the antigens in it. Table 10.3.1 shows how.

Table 10.3.1 Blood groups

Blood group	Red blood cell antigens	Plasma/serum antibodies
A	A	anti-B
B	B	anti-A
AB	A and B	no antibodies
O	no antigens	anti-A and anti-B

Rhesus

The rhesus factor is a protein that is either present or not present on the surface of the cells in your body. If you have the rhesus factor, then you are positive (e.g. A+). If you don't have it, then you are negative (e.g. AB–).

Forensic use of blood groups

Different racial groupings tend to have different blood groups and so the blood groups found at a crime scene might narrow the range of suspects. Table 10.3.2 shows how.

Table 10.3.2

Blood group	Proportion of Australians with that blood group	Racial grouping the blood most likely came from
A	39% (A+ = 32%, A– = 7%)	Aboriginal and Torres Strait Islander Peoples, Caucasian (European, North African, Middle Eastern)
B	11% (B+ = 9%, B– = 2%)	Asian, African
AB	4% (AB+ = 3%, AB– = 1%)	all racial groups
O	46% (O+ = 38%, O– = 8%)	all racial groups

Determining blood groups

If blood is mixed with any foreign antibodies, then the blood will coagulate (clump). Table 10.3.3 shows how each blood type reacts with different antibodies. The blood group of an unknown blood sample can therefore be determined by mixing it with different antibodies.

Table 10.3.3

Blood group	Mixed with anti-A antibody	Mixed with anti-B antibody
A	coagulates	nothing happens
B	nothing happens	coagulates
AB	coagulates	coagulates
O	nothing happens	nothing happens

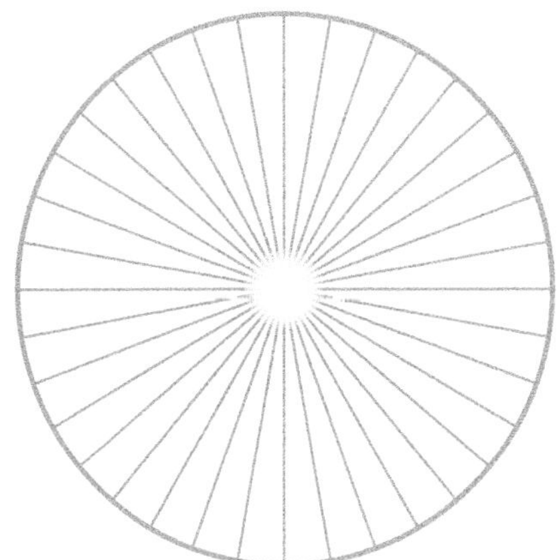

1. Use the circle on the left to construct a pie chart showing the occurrence of different blood groupings in Australia.

2. List the specific blood groups A+, A–, B+, B–, AB+, AB–, O+ and O– in order from most common to least common.

____

____

10.3 Blood typing

3 Define the following terms:

(a) rhesus factor ______________________

(b) coagulate. ______________________

4 Half of Australians are either group O or group AB. These blood groups are also common in all racial groups. Hence, a sample of blood group O or AB tells investigators very little. Despite this, all blood drops found at a crime scene will be tested for their blood group. Propose reasons why.

____

____

5 The proportion of people of blood group A+ is about the same as O+. Despite this, blood spatters of group A+ are far more useful to investigators than O+. Explain why.

____

____

6 A blood drop is found at a crime scene. Its blood group is the relatively rare AB–.

(a) State the percentage of the Australian population that has this grouping.

____

(b) Discuss whether this blood group would make it any easier to find the suspect.

____

7 Four samples of blood from a crime scene were mixed with two different antibodies. The results are shown below. Identify the blood group of each sample.

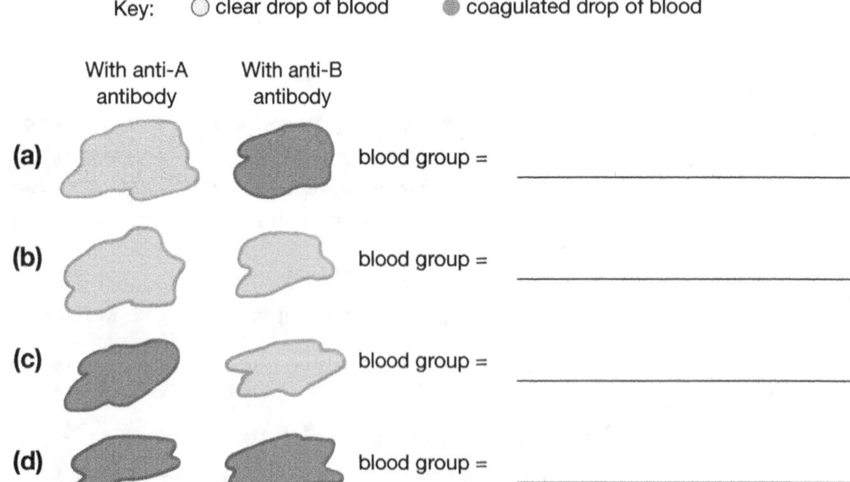

RATE MY UNDERSTANDING
Shade the face that shows your rating

10.4 Mapping your teeth

Science inquiry skills

FOUNDATION | STANDARD | ADVANCED

Processing & Analysing | Communicating

Dentists keep accurate dental charts that record any natural irregularities (differences) in their patients' teeth and the location and size of fillings, pins and crowns that have been put in at any time. Teeth are very tough and tend to outlast the rest of the body and provide a way of identifying the body of someone who has been missing for a long time. This is done by matching the teeth with dental charts. Below are the dental charts of an 'average' healthy child and an 'average' healthy adult.

fillings, pins, crowns (*n*) treatments made to the teeth by dentists

irregularities (*n*) parts that are not normal

outlast (*v*) to last longer than

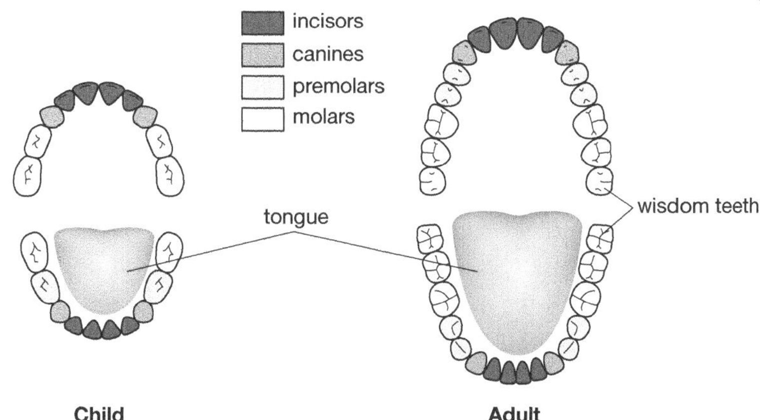

1. Compare the teeth of a child with those of an adult by completing the table on the right.

2. Why do you think children have fewer teeth than adults?

____

____

3. Run your finger over your teeth to count how many teeth you have. What is the total number?

____

Numbers of teeth	Child	Adult
Total number		
Incisors		
Canines		
Premolars		
Molars		
Wisdom teeth		

4. Based on the number of teeth you counted in your mouth, assess which of the above diagrams (child or adult) best describes your teeth.

____

5. Modify the diagram you chose in question 4 so that it more accurately records your teeth. Do this by:
   - crossing out any teeth you don't yet have
   - marking any teeth that are still in the process of coming down
   - crossing out any teeth that are missing for some reason
   - marking any teeth that have fillings, caps or other dental work
   - marking any teeth that have chips in them
   - marking any large gaps you have between your teeth.

6. Dental records won't always be an accurate way of identifying a body. Propose a reason why.

____

RATE MY UNDERSTANDING
Shade the face that shows your rating

10.5 DNA media analysis

Science understanding

FOUNDATION | **STANDARD** | ADVANCED

Read the article below about how DNA evidence helped prove a man was innocent and then answer the following questions.

conviction (*n*) decision that someone is guilty of a crime

exonerate (*v*) to find that someone is innocent

overturn (*v*) to reverse (a decision)

violation (*n*) breaking the law

Thirty years on, DNA tests free US man convicted of rape

A US man has tasted freedom for the first time in nearly 30 years after a judge overturned his conviction because DNA evidence showed he did not rape an 11-year-old girl.

'It finally happened, I've been waiting,' Raymond Towler, 52, said today as he hugged sobbing family members in the courtroom.

He walked from the courthouse, arms around relatives, amid the smell of freshly cut grass, blooming trees and a brisk wind off Lake Erie. He was headed to an 'everything on it' pizza party.

Towler had been serving a life sentence for the rape of a girl in a Cleveland park in 1981. Prosecutors received the test results Monday and immediately asked the court to free him.

In a brief, emotionally charged session, Cuyahoga County Common Pleas Court Judge Eileen Gallagher recapped the case, discussed the recently processed DNA evidence and threw out his conviction. She also told him that he can sue over his ordeal.

'You're free,' the judge said, leaving the bench to shake Towler's hand at the defence table. The judge choked back tears as she offered Towler a traditional Irish blessing.

The Ohio Innocence Project, an organisation that uses DNA evidence to clear people wrongfully convicted of crimes, said Towler was among the longest incarcerated people to be exonerated by DNA in US history.

The longest was a man freed in Florida in December after serving 35 years, according to the project.

Towler was arrested three weeks after the crime when a park ranger who had stopped him on a traffic violation noticed a resemblance with a suspect sketch. The victim and witnesses identified him from a photo, police said.

Carrie Wood, a staff attorney with the project, said the identifications were questionable.

The latest technology allowed separate DNA testing of a semen sample and other genetic material, possibly skin cells, she said.

'That was the test result that we got this week and it excluded Mr Towler,' she said. 'Because Mr Towler's conviction was in '81, the technology did not exist to do the kind of DNA testing that we can do now.'

Attorneys with the project at the University of Cincinnati have been working on the Towler case since 2004, and Towler said that and his faith had given him hope.

Prosecutor Bill Mason said his staff would test crime-scene evidence to try to identify the attacker.

Source: www.theage.com.au, 6 May 2010

10.5 DNA media analysis

1 How much time did Raymond Towler serve in prison before he was released?

2 What was the crime for which Towler was convicted (found guilty) and sent to prison?

3 Outline the evidence used originally to convict Towler.

4 How old was Towler when convicted?

5 List the samples that were collected from this crime scene from which DNA could be extracted.

6 Which one of those samples was used to release Towler from prison?

7 Explain why the samples were not originally tested for DNA.

8 Name the US-based organisation that uses DNA to clear people wrongfully convicted of crime.

9 How long had the project been working on this case?

10 What is the longest time spent by a person in prison in the USA before being released on DNA evidence?

11 Some people have suggested that all crimes should be re-assessed using new DNA technology. List the advantages and disadvantages of this proposal.

12 Assess if cases like this would ever have been reviewed without the development of fast computers.

RATE MY UNDERSTANDING
Shade the face that shows your rating

10.6 Poisons

Science understanding

FOUNDATION | **STANDARD** | ADVANCED

Throughout history, poisons were used as a murder weapon. This is because they were:

- relatively easy to get, many coming from common plants such as belladonna
- relatively easy to administer, most commonly slipped into food or drink
- difficult to detect since their symptoms resembled those of natural health problems such as dizziness, vomiting and diarrhoea. The murderer was therefore unlikely to be caught. If the poison was detected, by then the murderer was far away from the crime scene.

Some of the common poisons used in the past are shown in Table 10.6.1.

Table 10.6.1 Common poisons used in the past

	Arsenic	Cyanide	Deadly nightshade	Death cap mushroom	Strychnine
Found in	rocks	laurel plant	belladonna plant	a deadly form of mushroom	seeds of a tree
Symptoms	weak pulse, vomiting, intense saliva production, eventual multi-organ failure	dizziness, convulsions, asphyxiation (inability to get enough oxygen), unconsciousness, very red skin	pupil dilation, dry mouth, fever, hallucinations (seeing things), coma	stomach cramps, diarrhoea, vomiting, coma—symptoms disappear after a day but then keep returning	restlessness, muscular spasms that rip muscles from their ligaments
Affects	circulatory and digestive systems	respiration reduced, muscle and nerve cells die because tissues cannot take up oxygen	heart and lungs, causing paralysis	digestive system, causing liver failure	muscles, triggering spasms, paralyses lungs

The use of poisons as a murder weapon is now very rare. This is because forensic pathologists can detect the presence of even small doses of poison in a body. Poisonings nowadays are usually accidental, from the environment or deliberate acts of terrorism.

Environmental poisoning

In the 1970s, the Bangladesh government attempted to improve peoples' health by providing a clean, fresh water supply. An estimated 10 million tube wells were drilled to access underground water.

1 Research the drinking water of Bangladesh.

(a) Assess the quality of drinking water from the tube wells.

____

____

(b) Explain why the drinking water is of this quality.

____

____

____

____

10.6 Poisons

Accidental poisoning

The Union Carbide pesticide factory in Bohal, India was the scene of an accidental poisoning in 1984.

2 Research this incident.

(a) What poisonous chemicals were released as a result of the leak from the factory?

____

____

(b) How many people have been affected by this disaster?

____

Deliberate poisoning

administer (*v*) to give, to use on someone
belladonna (*n*) a type of plant
detect (*v*) to discover
leach (*v*) to dissolve slowly

Chemical and biological weapons are poisonous gases that are deliberately spread in the air to kill, usually in a war or terrorist attack. Some well-known examples are described below.

- Mustard gas, phosgene and chlorine gases were used in World War I (1914–18). About 100 000 soldiers were poisoned this way.
- Zyclon B (or cyclon B) was used in the concentration camps of Nazi Germany during World War II (1939–45). This cyanide-based gas was used to kill an estimated 1.2 million Jews and political prisoners.
- Mustard gas and sarin gas were used by Sadaam Hussein's army in 1988 to kill an estimated 10 000 Kurdish people in the north of Iraq.
- Sarin nerve gas was used by the Japanese terrorist group Aum Shinrikyo. Sarin was released into the Tokyo subway in 1995, killing 12 people.
- Anthrax spores were sent through the mail in 2001 to US government officials, killing five of those people who opened the letters.

3 In the past, poisons offered a potential murderer many 'advantages'. List the advantages.

____

4 Poisons are rarely used nowadays to kill. Explain why.

____

(5) Strychnine poisoning is very painful. Propose a reason why.

____

6 Use the information on this worksheet to complete the following table.

Where	Date(s)	Poison	Estimated deaths per year
Bangladesh			
Bhopal, India			
Trenches of World War I			
Nazi Germany			
Iraq			
Tokyo			
USA			

RATE MY UNDERSTANDING
Shade the face that shows your rating  

10.7 Art forgery

Science as a human endeavour

FOUNDATION | **STANDARD** | ADVANCED

Paintings by important artists sometimes sell for millions of dollars. So it is not surprising that they are sometimes forged. Sometimes simple checks can reveal that an artwork is fake.

artificially (*adv*) not occurring naturally, done by people
fake (*adj*) false, not true
forge (*v*) to copy something illegally
pigment (*n*) a substance used to make the colour in paint
varnish (*n*) a clear layer used to cover the paint

Forgery check 1

Each painting should have a provenance, which is a detailed history of its owners since it was painted and where and how it was purchased each time it changed hands. If there is no provenance or there are parts of the history missing, then the artwork is probably fake. Otherwise, it could be real but stolen.

Forgery check 2

Each artist has their own characteristic style, colours, pattern of brushstrokes and signature. Sometimes, a check of these features is all that is needed to prove an artwork as fake.

Non-destructive investigation techniques

If more evidence is needed, then the next step for the forensic investigator is to use simple non-destructive techniques that do not damage an artwork.

- A stereomicroscope shows detailed brushstrokes and can indicate whether the varnish over the paint has been artificially cracked to make the painting look older than it is.
- X-rays show whether cracks are just on the surface paint layers (suggesting fake ageing) or pass through all the paint layers to the canvas below (suggesting real ageing).
- Infrared (IR) radiation reveals painting and sketches under the surface image, which would indicate that the painting has been changed, painted over or had bits glued on.
- Ultraviolet (UV) radiation causes materials within the varnish and pigments that make up the paint to fluoresce (shine). The fluorescing colour can indicate the era in which the painting was made. For example, materials from the 1800s usually fluoresce blue-green.

Destructive investigation techniques

Destructive techniques of analysis are only used when all other techniques have failed to determine whether an artwork is fake or not. A small scraping of paint is taken from an edge or from a flake of paint that has come off. The destructive techniques commonly used are X-ray analysis and mass spectrometry.

- X-rays scatter (reflect) off different pigment crystals in the paint in different ways. Although the colours of the pigments may have faded and be unrecognisable, X-ray scattering will identify the exact pigment used. The invention of each new pigment is well documented and so a painting can be aged from the pigments found in it. For example, the presence of Prussian blue indicates that the artwork must have been painted after 1704.
- Mass spectrometry can determine the amount of impurities in the pigments. The purer the pigment is, the more recent it is. This technique can also determine whether the pigments were mixed with egg yolk, oil or water. For example, oil paints were only invented in the 15th century (1400s) and water-based paints (acrylics) in the 20th century (1900s).

1 Is it easier to forge paintings of artists who are dead than the paintings of those still alive? Provide reasons for your answer.

____

____

2 Define the following terms:

(a) provenance ______

(b) UV ______

(c) IR ______

(d) fluoresce. ______

3 A 30-year gap appears in the provenance of a painting by Picasso. Explain what this could mean. ______

4 What are two ways in which X-rays can assist forensic investigators in determining whether a painting is real or a fake? ______

5 An 'old' painting by Titian has cracks all across its varnish but X-rays show there are none in the paint itself. Determine whether the painting is real or a fake.

______

6 A Claude Monet painting of a street scene is tested with IR radiation. It shows that there is a sketch of a person's face on the canvas but no street scene. Explain why the painting is unlikely to really be by Monet.

______

7 Use the following information to classify the paintings as real or fake. Justify each of your classifications.

Painter	Lived	Painting tested by	Evidence found	Real or fake	Justification
von Guerard	1811–1901	mass spectrometry	used acrylics		
Rembrandt	1606–69	X-ray scattering	includes Prussian blue		
Jackson Pollock	1912–56	UV fluorescing	pigments fluoresce blue-green		
Ghirlandaio	1449–94	mass spectrometry	uses oils		
Tom Roberts	1856–1931	IR analysis	major features of a portrait were changed by painting over them		

RATE MY UNDERSTANDING
Shade the face that shows your rating

10.8 Faking meat and fish

Science understanding

FOUNDATION	STANDARD	ADVANCED

invaluable (*adj*) extremely useful
sashimi (*n*) a Japanese dish of bite-sized pieces of raw fish
substitute (*n*) replace something with a different item

To make bigger profits, unscrupulous people have long substituted expensive types of meat and fish with cheaper types. However, different animal species have different coding in their DNA. This makes DNA testing an invaluable tool to determine whether or not meat and fish are what they are claimed to be. Below are some ways in which DNA testing has proved produce to be 'fake'.

- In 2010, a fisherman tried to sell a tonne of mud crabs at the Sydney Fish Markets. DNA in the crabs proved that they were a Queensland species that is illegal to catch. This fisherman had instead claimed that they were legally caught in the Northern Territory.
- In 2010, an American undercover film crew found that samples of sashimi from restaurants in Seoul (South Korea) and in Los Angeles (USA) included whale meat instead of the fish normally used (such as tuna). The DNA coding of this whale meat was nearly identical to that from meat from sei whales bought in Japan in 2007 and 2008. Although whale meat is safe for human consumption, it is claimed that whales caught by Japan are for scientific research and not for food.
- In 2013, DNA testing in the United Kingdom and Ireland showed that the 'beef' in popular brands of frozen lasagne and in frozen hamburger patties contained horse meat. Sometimes there was no beef at all! Although horsemeat is safe and eaten in some European countries, it is very unpopular in the UK and Ireland and consumers would react badly if they knew their 'beef' was actually horse.
- In 2013, pork DNA was found in pies prepared in Ireland for prison kitchens there and in England. Due to religious dietary rules, Muslims and Jews do not eat pork. The findings resulted in a total of 3663 products from the pie manufacturer being withdrawn from sale.

1 Propose reasons why the catching of some species of mud crabs might be illegal in some states but not in others.

____

____

2 **(a)** State the probable source of whale meat found in the sashimi described above.

____

(b) Why do you think the film crew needed to collect their sashimi samples while undercover?

____

____

3 Propose reasons why horse meat would be substituted for beef in frozen lasagne and hamburger patties.

____

____

10.8 Faking meat and fish

4 **(a)** Explain why pork is not eaten by Jews or Muslims.

(b) The 'beef' pies that contained pork were destined for prisons. Assess whether society should be concerned with the religious customs and other dietary needs of prisoners.

(c) Justify your response.

5 McDonald's sells beef burgers around the world but not in India. Hindus do not eat beef so McDonald's sell chicken burgers and vegetarian dishes instead in India. Explain why it would be nearly impossible for McDonald's to 'fake' their chicken burgers and include some beef.

RATE MY UNDERSTANDING
Shade the face that shows your rating

10.9 Literacy review

Science understanding

FOUNDATION	STANDARD	ADVANCED

1 Use the clues to identify the words.

Clue	Word
The scientific study of evidence	f _ r _ _ _ _ _ _
Deliberate fire	a _ s _ _
The killing of another person	h _ _ _ _ _ _ _
The study of guns and bullets	b _ l _ _ _ _ _ _ _
Person who sees a crime	e _ _ _ _ _ _ _ _ _
The systematic dissection of a dead body	a _ _ _ _ _ y
Dead body	c _ _ _ _ _
A chemical 'fingerprint' found in every cell in your body	D _ _
Absorbs water and oil	p _ _ _ _ _
Microscopic organisms found in water	d _ _ _ _ _ _
The large bone in your upper leg	f _ _ _ _
Loops, whorls, arches and composites are different types	f _ _ _ _ _ p _ _ _ _ _
The use of a liquid to separate colours in dyes	c h _ _ _ _ _ _ _ _ _ _ _ _
Raised printing on a banknote	i _ _ _ _ _ _ _
The coloured muscular ring at the front of the eye	i _ _ _
The back of the eye	r _ _ _ _ _
Liquid that dissolves solids and other liquids	s _ _ _ _ _ _
A technique that matches parts of the face to reconstruct the face of a criminal	i _ _ _ _ _ _ k _ _
Plastic film used to make Australian banknotes	p _ _ y _ _ _
A type of fingerprint	w _ _ _ l

2 Once you have identified the words in question 1, find them and highlight them in the word find below.

S	R	H	I	L	P	I	S	X	L	T	O	C	I	S
T	A	N	I	T	E	R	D	F	N	F	D	N	F	M
N	E	Y	E	W	I	T	N	E	S	S	T	O	O	O
I	T	H	I	B	S	R	V	T	N	A	L	S	R	T
R	K	P	I	A	K	L	Y	P	G	T	C	R	E	A
P	R	H	C	C	O	K	R	L	J	I	I	A	N	I
R	C	O	T	S	M	M	I	O	T	I	I	K	S	D
E	D	I	C	I	M	O	H	S	H	C	R	S	I	E
G	N	G	N	Y	D	B	I	P	T	W	Y	I	C	T
N	A	J	C	K	S	L	F	E	M	U	R	F	S	P
I	B	R	Y	O	L	P	O	L	Y	M	E	R	O	S
F	Y	H	P	A	R	G	O	T	A	M	O	R	H	C
O	T	M	B	X	O	P	C	T	J	L	O	O	N	F
B	Q	T	K	C	D	T	S	Y	U	U	C	R	Y	J
W	B	U	F	S	W	P	F	E	S	A	S	B	Q	I

RATE MY UNDERSTANDING
Shade the face that shows your rating

10.10 Thinking about my learning

1 Study the image of a crime scene. Based on the image, list the forensic techniques that you have learned that could be used to help solve this case. For each technique listed, explain what information could be gained.

2 Advances in technology have allowed more accurate detection of people who are using false identities and have helped in preventing identity fraud. Describe three technologies that have helped in this field.

3 Describe one interesting thing you have learned in this unit that you haven't mentioned in the first two questions.

4 State one topic you would like to know more about from this chapter.

5 State one question that you still have about forensic science.